Energy and Environmental Justice

"Tristan Partridge's *Energy and Environmental Justice* is a well-written, essential primer of how energy and environmental justice are linked through systems, movements and power. Organized through theories, concrete examples from throughout the world, and research agendas, Partridge's synthesis is incredibly important. The book is short and readable, and re-centers energy through work and politics, foregrounds indigenous, racial and gender perspectives, and usefully explains what justice, transition, and degrowth means grounded in everyday struggles. This book should be used in many classrooms and organizing spaces."
    —Julie Sze, author of *Environmental Justice in a Moment of Danger*

"I couldn't put this book down. *Energy and Environmental Justice* is a powerful treatise and meditation on the necessity of taking energy justice seriously, linking it to generations of scholarship and activism by researchers and advocates of environmental justice, Indigenous Studies and Indigenous sovereignty (among other fields). One of the most theoretically rich and methodologically challenging books I have come across in a long time, *Energy and Environmental Justice* has forced me to completely rethink energy justice from the ground up. It also leaves me hopeful because Partridge articulates the importance of pushing beyond modes of mere resistance to enact strategies of redefinition, refusal, and transformation in how we think about and engage with energy, labor, systems of oppression, multispecies relationships, and so much more. Tristan Partridge has produced a highly original volume that will breathe new life into the field and will set the tone for the next generation of scholars."
    —David N. Pellow, author of *What Is Critical Environmental Justice?*

"This insightful work takes a comprehensive look at how energy production and consumption intersect with the global movement for environmental justice, illuminating both the socioecological catastrophe of the status quo and the possibilities for more just and sustainable energy relations. This book should be a first stop for readers seeking to understand and create the energy systems of the future."
    —Anna J. Willow, author of *Understanding ExtrACTIVISM: Culture and Power in Natural Resource Disputes*

"This book re-politicizes debates about desired energy futures, rooting those debates in the demands of diverse social movements fighting for radical social change."

—Federico Demaria, coeditor of *Pluriverse: A Post-Development Dictionary* and *Degrowth: A Vocabulary for a New Era*

"This accessible and concise handbook should be required reading for every student in environmental studies and related fields. It helpfully reconnects energy research with the radical perspectives, activist roots, Indigenous insights, and key concepts required for building the future we need, clarifying what just transitions require. Its synthesis of ideas, methodologies, and recommendations for a critical energy justice research agenda are valuable for scholars who wish to be accomplices in community struggles that refuse oppression and generate new relationships."

—Corrie Grosse, author of *Working across Lines: Resisting Extreme Energy Extraction*

Tristan Partridge

Energy and Environmental Justice

Movements, Solidarities, and Critical Connections

Tristan Partridge
Institute for Social, Behavioral and Economic Research
University of California
Santa Barbara, CA, USA

ISBN 978-3-031-09759-1   ISBN 978-3-031-09760-7   (eBook)
https://doi.org/10.1007/978-3-031-09760-7

This Palgrave Macmillan imprint is published by the registered company Springer Nature Switzerland AG.
The registered company address is: Gewerbestrasse 11, 6330 Cham, Switzerland

ACKNOWLEDGMENTS

This book project took shape while I was working at the ICTA Institute of Environmental Science and Technology at the Universitat Autònoma de Barcelona. Particular thanks go to Sofia Avila for inspiration and collaboration over a number of years—your work, individually and collectively with others, is vital for everyone who wants to bring environmental justice approaches and insights to bear on contemporary energy research. I am most grateful for your dedication and friendship.

I wholeheartedly thank all the staff and participants who made our workshop possible at ICTA on Energy & Environmental Justice (October 2019). Everyone who attended (in person and remotely)—together with everyone who followed up through shared texts and further exchanges—has influenced my work. I look forward to continuing these conversations. I am greatly indebted to Dana Powell and David Pellow—thank you for your invaluable inspiration—and to Anna Willow, Sourayan Mookerjea, and especially Javiera Barandiarán—my most gracious thanks to all of you for your guidance and support at different stages of the planning, researching, and writing processes. Many other people have been very generous with their time while they continue to do incredible world-changing work: thank you to Corrie Grosse, Winona LaDuke and Honor The Earth, annelise lewallen, Kyle Powys Whyte, Joan Martínez-Alier, Dustin Mulvaney, Hamza Hamouchene and The Transnational Institute, Ruth Santiago, Hilda Lloréns, Catalina de Onís, Sergio Tirado Herrero and L'alianca contra la Pobreza Energètica; Christos Zografos, Federico Demaria, Roberto Cantoni, Mariana Walter, Rishabh Raghavan, Silvia Pergetti,

Jia-Ching Chen, Erin Goodling, Julie Sze, Elia Apostolopoulou, Stacia Ryder, Lucrecia Wagner, Brototi Roy, Daniela del Bene, the EJAtlas team, Sergio Villamàyor-Tomas, the Degrowth Reading Group, Giorgos Kallis, Sonia Graham, Santiago Gorostiza, Anke Schaffartzik; and everyone I met and connected with through ICTA.

Our Energy & EJ Workshop—as well as the fieldwork in India I conducted on the Power Farmers solar program in Uttarakhand and the SPICE solar cooperative in Gujarat in 2018–19—were facilitated thanks to financial support from the Spanish Ministry of Science, Innovation and Universities through the "María de Maeztu" program for Units of Excellence (MDM-2015-0552). Work on oil extraction and environmental justice measures in coastal California was partially supported by a Collaborative Research Initiative Grant from the Institute for Social, Behavioral and Economic Research at the University of California, Santa Barbara. My thanks also to Rachael Ballard at Palgrave for seeing this project through to publication, to Elizabeth Graeber for introductions, to Rubina Infanta Rani for project coordination, and to the anonymous reviewers whose reflections on this work helped generate a number of improvements.

Over recent years, it has been an honor to connect with so many wonderful colleagues and friends working on issues of energy and justice. Thank you to Julie Maldonado, Kirk Jalbert, Stephanie Paladino and members of the ExtrACTION topical interest group at the Society for Applied Anthropology (USA); Barbara Herr Harthorn for all kinds of support; Justin Kenrick, Arturo Escobar, Alberto Acosta; Lisa Sun-Hee Park, Mona Damluji, Brandon Fastman, Bregje van Veelen, Casey Walsh, Emiko Saldivar, Bishnupriya Ghosh, Anabel Ford, Charlie Hale, Alice O'Connor, Emily Roehl, Stephan Miescher, Janet Walker, Ry Brennan, Jéssica Malinalli Coyotecatl Contreras, Ingrid Feeney, Emily Williams, Theo LeQuesne, Bhavna Shamasunder, Lane Clark, Dena Montague, Dave Novak, Ayumi Nakamura, and all the participants at the Energy Justice in Global Perspective Sawyer Seminar at UC Santa Barbara (especially the Re-Centering Energy Justice Symposium, 2019); Jamie Cross, Sheila Davis, and everyone who participated at the Capitalising on the Sun event in Edinburgh; the Antipode Geographies of Justice team; Ana Stojilovska, Michael LaBelle and the Energy Policy Research Group at CEU (who hosted my talk "Rethinking energy justice: Alternative histories and possible futures"); James McCarthy, Noel Healy, and attendees at the Clark University Graduate School of Geography seminar (where Sofía Avila and

I presented some of this work under the title, "Energy justice: politics beyond policies").

Through fieldwork trips, residencies, and conferences, I have been very fortunate to engage with people whose committed and incisive work is a key source of motivation. My thanks to David Harvey and ICSI seminar participants at The New School; Lynn Badia, Graeme Macdonald, Cymene Howe, Bob Johnson, Dominic Boyer, Annabel Pinker, and friends made through the Petrocultures network (especially in Newfoundland and Glasgow); Roman Sidortsov, Sara Fuller, Tony Reames, Stefan Bouzarovski; Jeffrey Jacquet, Julia Haggerty, Anne Junod, and the Energy Impacts Symposium team; attendees at the 2018 International Conference on Advanced Materials, Energy & Environmental Sustainability (and everyone who arranged my invitation from UPES Dehradun, India); Valerie Kuan, Shawn Barcelona, and Stuart Tyson Smith at ISBER (UCSB); Francesca Bray, Maya Mayblin, Gaia von Hatzfeldt, Lucy Lowe; Fellows and friends at the American Academy in Berlin (Spring 2022); my dear colleagues on multiple collaborative projects, including Off The Grid: Relational Infrastructures for Fragile Futures (University of Edinburgh); Knowing The Underground, and the CREW Center for Restorative Environmental Work (UCSB); and Energy, Risk and Urgency (connecting across the Universities of California and Cardiff); Porfirio Allauca in Cotopaxi; Carlos Quinto Cedeño Bermeo in Manabí; Fudabhai Parmar and all members of the Dhundi Saur Urja Utpadak Sahakari Mandali (Solar Cooperative); Neha Durga, Tushaar Shah, Shilp Verma and everyone at IWMI in Anand; Votan Ik; Greg Carr; The Isle of Eigg Heritage Trust; Bharat Dogra for insights on EJ issues across India; and Dr. Anupam Bhandari for many years of friendship and fieldwork support in Uttarakhand.

CONTENTS

LIST OF IMAGES

CHAPTER 1

Energy from the Perspective of Environmental Justice

Abstract The struggles of diverse EJ and Indigenous movements reaffirm that other worlds are possible. This section highlights how, emergent within these global struggles, there are insights and strategies for reshaping energy research on the unjust dimensions of contemporary energy systems. Studying energy from EJ perspectives further involves engaging with discrimination and the distribution of power, in all its forms, across societies. This section details conceptual, historical, and theoretical approaches that reassess energy's role in producing injustices, drawing in particular on (i) critiques of extractivism, anti-politics, racial capitalism, settler colonialism, and the political implications of energy when understood as the capacity to do commodifiable work, and (ii) closer engagement with autonomous, prefigurative, and sovereignty-based movements. These repoliticized perspectives on energy underline how multiple strands of EJ inquiry and action come together to identify, resist, and resolve environmental inequalities.

Keywords Energy histories • Extractivism • Labor exploitation • Racial capitalism • Settler colonialism • Autonomy

© The Author(s), under exclusive license to Springer Nature Switzerland AG 2022
T. Partridge, *Energy and Environmental Justice*,
https://doi.org/10.1007/978-3-031-09760-7_1

Introduction: EJ Movements and Critiques

The expansion of modern energy systems is a leading cause of environmental change and conflict around the world. Members of the Siekopai, Waorani, Siona, Kichwa, Shuar, and Cofán Indigenous nationalities in Ecuador, together with their neighbors and animal relations in the Amazon, have seen the forests and rivers they depend on become toxic as a result of decades of oil extraction. Campaigners in southern Algeria are confronting water pollution and social conflicts created by the corporate exploitation of shale gas. Farmers, villagers, and Adivasi communities across Jharkhand—one of the poorest states in India and home to the country's largest coal mines—continue to fight the loss of land and air pollution caused by fossil fuel extraction and energy generation. Social movements the world over have mobilized to address such impacts, but resistance itself carries risks. The killing of Environmental Human Rights Defenders, including anti-dam activists, is becoming an 'epidemic' in some countries (Del Bene et al. 2018). Native American communities and allies who gathered at Standing Rock in 2016 to stop the massive Dakota Access oil pipeline were met with militarized repression and police violence. All the while, burning fossil fuels for energy is a primary driver of global climate change. The scope of these many different conflicts and changes is planetary, but marginalized communities and bodies disproportionately suffer the consequences. That is the definition of environmental injustice—and an inseparable aspect of contemporary energy relations.

The term 'energy relations' refers to both the material infrastructures that constitute energy systems as well as the socioecological relationships that are variously generated, supported, destroyed, or devalued through the operation of those systems. Energy is itself a social relation. Over millennia, societies have developed different methods of harnessing, sharing, and using energy. These energy practices foster differing lifestyles and political practices, shaping the spaces where people live and work, and altering how people relate to the Earth's ecosystems. Today, energy use is at the heart of economic and industrial activity, channeled into a widening range of personal, productive, and profit-driven endeavors and creating an array of social and environmental impacts. This has long been the case: from the clearing of medieval forests for ore smelting and boat manufacture, to the enslavement of peoples and theft of natural wealth throughout colonialism, to the standardization of work and an accompanying explosion in greenhouse gas emissions during European capitalist industrialization. Today it continues through an era of neoliberalism characterized by a rampant disregard for the consequences of over-consumption: half of all global CO2 emissions since 1751 have been created after 1990 (IEEP 2020). Acknowledging the leading work of Native struggles for energy justice (e.g., LaDuke 1999; EJNA 2009) and justice-oriented energy activism (e.g., Sze 2005), this book examines how energy practices and their impacts relate to environmental change, community activism, and injustice. To do so is to examine energy from the perspective of environmental justice.

Environmental justice (EJ) work addresses how environmental change relates to other injustices—including racism, sexism, ableism, ethnocentrism, and settler colonialism—which occur when some people benefit by systematically inflicting harms and risks on others (Whyte 2017a). Injustices are systematic when those who benefit from such violence, and those who suffer it, are separated across social categories of difference— such as race, gender, sexuality, ability, class, and species—categories which are often created and imposed on marginalized groups by more powerful sectors of society as a way of furthering their interests (Whyte 2017a; Pellow 2018). The influence of these systems of inequality is witnessed in multiple different forms of systemic violence and injustice. Studying energy from an EJ perspective therefore involves not only addressing uneven experiences of environmental harms that result from energy practices—such as the contamination of air and water, loss of land and wildlife, greenhouse gas emissions, and the toxification of habitats for both humans and nonhumans—it also means engaging with discrimination and the distribution of power, in all its forms, across societies.

Environmental justice activism and scholarship stems from the Civil Rights Movement in the US and grassroots struggles led by African American communities linking environmentalism with (anti)racism (Bullard 1990). Focused initially on how Black communities disproportionately suffer the ill effects of hazardous waste and industrial operations, EJ highlights how toxicity and environmental racism contribute to the violence inflicted upon communities of color by dominant sectors of society. Building momentum through the 1980s, the EJ Movement in the US drew upon direct action, civil disobedience, anti-racism organizing, and engaged academic research, as well as parallel struggles for justice within the labor movement and Native American communities (Cole and Foster 2001). EJ has since emerged globally as a field of study and focal point for activism, taking root in localized struggles against environmental degradation.

The 'environmentalism of the poor' refers to these parallel movements mobilized by marginalized and low-income communities in order to defend the living environments on which they depend (Guha and Martinez-Alier 1997). Together with diverse Indigenous struggles, this array of politicized environmental actions forms a global environmental justice movement: a 'movement of movements' partially connected by common goals, frames, and forms of mobilization, and sometimes confronting the same powerful adversaries (Martinez-Alier et al. 2016; Roy 2022). The Climate Justice movement similarly connects diverse local struggles which fight against core drivers of climate change (such as dispossession, exploitation, contamination, and industrial expansion) and also highlight the uneven global impacts of climate change—especially across the geographical and political South and with particular consequences for the sovereignty and wellbeing of Indigenous groups (Tsosie 2007; Building Bridges Collective 2010). Together, the many strands of EJ inquiry and action seek to identify, resist, and resolve environmental inequalities.

EJ is an expansive and expanding field, engaging with different accounts of how environmental justice is defined, promoted, challenged, disrupted, demanded, or defended. This book is a compilation of particular examples of 'EJ work' and 'EJ perspectives'—drawn from diverse sources including scholars, activists, critical thinkers, and community groups—that speak directly to energy-related injustices and concerns. The collected works articulate critical perspectives that have, to date, received relatively little attention in the broader field of energy research, energy studies, and 'energy justice.' Energy justice scholarship is now an established body of work that considers not only the technical and economic dimensions of energy systems but also their connections to questions of ethics, politics, and justice. Much of the energy justice literature, however, glosses over its relation with EJ scholarship and activism. Some very visible works of energy justice go as far as to disavow the activist and Indigenous roots of the term (e.g. Jenkins 2018). In direct contrast, this book highlights how, emergent within the struggles fought by the global movement of EJ movements, there is a wealth of knowledge, strategies, and insights that hold vital lessons both for energy research and for activism that specifically addresses the unjust dimensions of contemporary energy systems.

The rest of Chap. 1 details conceptual and theoretical approaches that reorient how we think about energy, reconnecting energy research with a range of radical, critical, and EJ perspectives. These multiple fields of research and action include critiques of extractivism, anti-politics, racial capitalism, settler colonialism, and the political implications of energy when it is understood as the capacity to do commodifiable work. Repoliticizing energy research means re-engaging with these critiques and with autonomous, prefigurative, and sovereignty-based movements.

Chapter 2, "Transitions Beyond Crisis," explores the vital roles that EJ movements and analyses play in building future energy systems that are more equitable, less destructive, and which support a just transition away from global dependency on fossil fuels. As the effects of climate change continue to generate extreme weather events and to devastate many communities who have contributed least to global greenhouse gas emissions, the importance of such a transition is undeniable. With notable exceptions among certain interest groups and populist leaders, the urgency of taking action in order to change course is widely accepted. However, Indigenous and EJ scholars caution against the uncritical pursuit of urgent action—that is, without also directly addressing social structures and conditions that create injustices in the first place.

The uncritical approach is seen when governments and other actors use the urgency of injustice—the immediate need to halt acts of violence and environmental harm and to support those who have suffered—as a way to justify policy changes that are still rooted in current systemic inequalities (and depend on their perpetuation) (Partridge et al. 2018). A more comprehensive approach additionally attends to concurrent concerns including increasingly exploitative and precarious labor relations (Newell and Mulvaney 2013); the prioritization of corporate and state interests and obeisance to the power held by those institutions (Pulido 2017); and the growth imperative of most economies (Kallis et al. 2020). More fundamentally still, critical approaches ask whether consent, trust, accountability, and reciprocity underpin the relationships that are being mobilized within responses to environmental injustice, particularly in connections between Indigenous peoples, governments, corporations, and other societies (Whyte 2020). Such engaged approaches to building energy futures recast transitions as collaborative projects of justice, building in particular on diverse movements for restorative environmental justice.

Chapter 3 maps out a Critical Energy Research agenda, drawing together the radical and transformative approaches covered in Chaps. 1 and 2. Critical energy research applies EJ perspectives and movement-generated insights to energy studies, analyzing energy relations in light of their role within broader systems of (re)production, politics, economics, and ecological regeneration. Reorienting our work in these ways carries implications that are not only theoretical and political but also methodological. Accordingly, critical energy research also reconsiders the roles and responsibilities of those who write about injustice, calling for renewed and committed forms of reflexivity within energy studies.

REPOLITICIZING ENERGY HISTORIES

EJ perspectives highlight how the history of energy emerges within the logics of exploitation and extraction that drove systems of chattel slavery and imperialism from the fifteenth century onwards and which continue to fuel anti-Blackness and (neo)colonialism today (Yusoff 2018; Bledsoe and Wright 2019; Fiori 2020). Tracing the role of energy extraction and transfer within histories of exploitation shows how dominant ideas about energy (and the material organization of energy systems) are inseparable from—and are used to justify—inequality, exclusion, and the devaluing of others. The establishment and growth of particular energy regimes continue to shape geographies of displacement and dispossession (McKittrick 2006). The scales on which these patterns operate show no signs of shrinking. Life is extracted from certain regions where growing numbers of people are condemned to live in uninhabitable places, their homes effectively rendered inhuman rather than human spaces (McKittrick 2013). Energy histories, therefore, need to recognize how energy relations play a central role in processes of racialization, understood as a "conglomerate of sociopolitical relations" that "discipline humanity into full humans, not-quite-humans, and nonhumans" (Weheliye 2014, p. 3). This means scrutinizing not only the kinds and amounts of energy being used—as well as how energy is generated or acquired, exploited, and put to use (and for whose benefit and at whose cost)—but also how energy relations themselves *produce* multiple systemic injustices. EJ perspectives ensure we recognize that energy histories are also histories of exploitation.

Violent racial ideologies that constructed Indigenous people and Black people as less-than-human fueled the murder and dispossession of Native peoples and the institutionalization of the Trans-Atlantic Slave Trade as the first 'industrial-scale energy infrastructure' (Lennon 2017). Coercion and the exploitation of African labor in the Caribbean sugar plantation economy—the rapid extraction of profit from soil and slave labor—was integral to the growth of industrial capitalism and Europe's period of geopolitical dominance (Carrington 2003; Mahmud 2013; Williams 2014

[1944]). Within this emergent global economic system, slavery and the sugar industry solidified a 'cultural understanding of production' based on maximizing profit from the transformation of energy and materials (in whatever form or location they were to be found), no matter what the human or environmental cost (Hughes 2017). Those maximizing principles have persisted across multiple energy regimes and epochs:

> As sugar plantations declined in economic significance, a new plantation complex emerged that fed slave-grown cotton from the United States to the massive textile mills in England, confounding the assumption that the two energy regimes—the slave-driven and the fossil-fueled—were developmental phases in conflict with each other. At the outbreak of the US Civil War, Liverpool was the greatest center of support for the Confederacy outside the American South. Even before it directly embraced coal and oil, then, the US was being shaped by a revolution in energy that did not so much shift the world from one regime to another as stack new energy sources on top of existing ones, intensifying and altering but by no means replacing the plantation complex. (Ziser et al. 2020, p. 549)

Subsequent energy infrastructures of different kinds have been constructed according to similar principles—sharing designs that support particular patterns of political and economic power.

Recognizing these continuities is one way to ensure that energy research moves beyond operating only in a diagnostic mode that merely identifies individual cases of energy-related injustice and adds them to the scholarly record. Critiquing the evident, contemporary injustices of an industry or field of activity is a partial approach that is only fully realized by also (re) historicizing the power relations that generate and sustain those endeavors (Trouillot 2003). Any critique of destructive energy operations thus remains incomplete without also historicizing the productive relations (and discourses of economic growth) that both drive global energy demand and account for most global energy use.

This 'unveiling' of the productive relations that constitute commodities is a means to more clearly identifying the forces and materials that make up energy systems; it also serves to expose previously hidden injustices, to challenge the politics of consumption, and to identify new possibilities for social change (Mulvaney 2019, p. 41). Whether injustices are 'hidden' is of course a question of perspective: they are often all too acutely evident to the people suffering their effects. Sometimes, however, the devastating, toxic impacts of particular industrial processes take many years to manifest—a time-lag, coupled with state deregulation, that polluting companies have willfully exploited in order to continue their damaging operations without detection (Pellow 2001). When environmental injustices *are* deliberately hidden by those responsible, the effect is to enable patterns of exploitation to continue. Following these dynamics through the "US energy story," Shalanda Baker traces that country's energy history as one that is "filled with injustice and with lives cut way too short by structural inequality":

> Our people—scattered in communities along the road that stretches from Port Arthur, Texas, to Lake Charles, Louisiana—were the enslaved and colonized. Perhaps it is no mistake then that the energy systems born out of that same soil produces so much wealth in the United States and simultaneously makes life nearly unbearable for Black and Brown bodies. The energy system has, in many ways, swapped out one system of extraction—legalized slavery—and replaced it with a more modern one, where oppression does not live in the lash, but in the toxic molecules that pollute our communities in higher numbers, wedge into our airways and waterways, and kill us. (Baker 2021, p. 26)

Logics of extraction and extractivism (see below) continue to define the majority of global energy infrastructures: systems of production based upon well-known, well-documented, and fatal forms of environmental destruction. Energy histories—when we see that they involve much more than a series of shifts from one predominant fuel source to another—reflect the deliberate infliction of the effects of contamination upon certain peoples, places, bodies, and communities. Generating more complete accounts of how and why injustices emerged in the past is another active mode of working towards justice in the present and into the future. Doing so requires that we rearticulate how the operation of energy systems connects with political processes and imbalances of power.

'Energy' is commonly used as an umbrella term to refer to various phenomena experienced as fuel, movement, heat, light, and power in different forms. Some energy sources we encounter in the natural world: the movement of wind and water; the heat of the sun or a wood-burning fire; or the toxic concentration of fossilized life that exists today as coal, oil, and methane gas. Other forms of energy, particularly electricity, are invisible (and invisibilized) but nevertheless permeate our interactions with the world around us. This immateriality created novel legal questions in the earlier social life of electricity: if it is not an object, can it be stolen? (Maris 1991). Yet the sheer pervasiveness of energy also makes it a difficult focus for study—increasingly taken for granted, embedded in everyday life for billions of people, and deeply enmeshed in overlapping social, legal, political and environmental entanglements (Appel 2012; Loloum et al. 2021). Despite energy's connections with most aspects of society—and the fundamental roles that energy relations play in (re)producing social life— energy systems (and much contemporary energy research) have become anti-political.

James Ferguson describes the anti-politics of the development sector operating as a bureaucratic-discursive 'machine' that creates institutional and ideological effects: reinforcing and expanding state (and corporate) power, systematically misrecognizing the lives of marginalized people, and depoliticizing social problems by reducing them to technical challenges (Ferguson 1990, 1999). Energy systems, immersed in their own anti-politics, tend to recreate all of those effects, not least because the energy sector and the development sector have become so tightly enmeshed.

For example, number seven of the 17 Sustainable Development Goals— part of the United Nations 2030 Agenda for Sustainable Development—is to "ensure access to affordable, reliable, sustainable and modern energy for all" (UN 2016). As Munro et al. (2017) point out, drawing on their work in Sierra Leone, while this goal clearly states its distributive objective ("energy for all"), there is relative silence on how that access is to be achieved and, crucially, on who is to take up roles of leadership, control, and ownership of the different forms of (energy, political, economic, social) infrastructure that will be required to deliver on that goal. SDG7 therefore "frames energy poverty in primarily technical-managerial terms, obscuring the political-economic dynamics of which it is a product" (Munro et al. 2017, p. 635). This recreates each of the aspects of anti-politics described above: assuming those who already hold positions of power (and control over energy systems) will, and should, continue to do so; misrecognizing how political-economic histories have *produced* the conditions of poverty that the Goals seek to address; and rendering experiences of marginalization as problems that could be resolved through technological innovation and expansion.

Against the trend among policymakers (and many scholars) of endorsing such depoliticized framings and 'techno-optimistic' energy solutions to societal problems, approaches to energy research that are more critical seek instead to *re-politicize* energy and energy justice (Healy et al. 2019). This does not mean, as is frequently understood by the phrase 'become politicized,' subjecting any proposal that is designed to address energy-related injustices to baseless review in the theatre of party politics (characterized primarily by entrenched modes of thinking and short-term points-scoring efforts between rival political factions). Instead, (re)politicizing energy issues means moving beyond programmatic models of justice and policy management in order to reconnect energy-related injustices to the political-economic dynamics that produced them.

Each of the following sections outline overlapping approaches for (re) politicizing energy research, drawing on EJ work on racial production; energy and the politics of work; extractivism; settler colonialism; and autonomy from the state. These different approaches to repoliticizing energy histories each involves a process of more carefully accounting for the co-emergence of certain energy relations and political-economic systems.

The physical capacities enabled by different energy sources have dramatically shaped productive, economic, and social dynamics in societies over time. It is then, perhaps, no surprise that energy systems and energy use are pervasively woven through notions of development. Tracing these changes, an historical view of 'energetic determinism' is common—plotting history through types and quantities of energy use and identifying transitions from one predominant mode to another, such as the shift during eighteenth century industrialization from dependence on biomass and bodily power to fossil fuels and fuel-consuming engines (Smil 2004, p. 554). Even though few today would defend Leslie White's evolutionary view of 'cultural development' as a function of increased energy production and consumption (1959), epochal studies play an important role by tracing how changes in energy practices foster other, sociocultural transformations (Nye 1998; Mitcham and Rolston 2013). For example, as Mitchell (2011) and others have shown, certain energy sources (especially fossil fuels) have directly influenced the formation, negotiation, and relative fragility or persistence of certain political systems (including modern democracy).

Accounts that rely too heavily on deterministic explanations, however, tend to obscure influential cultural and political dynamics—failing to account, for example, for the unequal ways in which energy use and well-being are related (both within and between countries) and the unevenness in how people experience those sociocultural transformations. Attentive to those injustices, and engaged in identifying both their impacts and their emergence, EJ research and activism underscore a vital starting point for critical energy research. That is, to show how environmental and energy-related injustices do not simply occur; rather, they are produced—by practices of accumulation and exploitation that regard some lives as mattering more than others. Uneven exposure to environmental benefits and harm is "often not accidental and unintentional, but rather a product of a particular way of organizing production and its constitutive social relations" (Newell and Mulvaney 2013, p. 133). Those productive and societal relations reflect (and reinforce) the dominant patterns of racism, sexism, ableism, ethnocentrism, and settler colonialism that operate across society.

RACIAL PRODUCTION

Critical EJ perspectives on energy scrutinize not only racial outcomes (and unequal experiences of the impacts of energy systems) but also racial production: the study of how those unequal experiences come about as inseparable aspects of the 'racialized production of differential value' (Pulido 2017, p. 528). This approach examines links between energy practices and multiple systems of domination that operate through the enforcement of structural inequalities, including ongoing histories of racial capitalism and settler colonialism (Kojola and Pellow 2021). Both are structures of violence that connect with (and uphold) the global dominance of extractive energy operations. In studying these relationships, EJ research articulates critical views on the state as a discriminatory site of conflict and also—the focus of this section—illustrates how racism underpins energy-intensive capitalist systems of production.

Theories of racial capitalism show how categories of racial difference are fundamental to the structural logic of contemporary capitalism and related economic systems including colonization and chattel slavery (Robinson 1983; Pulido 2017). Capital accumulation depends upon the transformation of places and practices into property and the concentration of ownership of such property among a relatively small number of people. Accumulation thus produces relations of inequality of many different kinds, including the separation of people who 'own' and manipulate land and resources for their own purposes from people who have been displaced and dispossessed (Melamed 2015). In addition to economic gain, this creates environmental privileges (such as access to clean air and water) for those who benefit from such control over human and nonhuman others (Park and Pellow 2011). These multiple relations of inequality are further reflected in the individualization and atomization of social life—processes driven by a rhetoric of individual responsibility that actually obscures or ignores how individuals are differently positioned, particularly in terms of access to economic stability and exposure to environmental harms.

Notions of difference are not only *exacerbated* within processes and practices of racial capitalism; rather, notions of difference are *weaponized* directly in order to facilitate different forms of accumulation. Racism both generates and thrives upon the physical and symbolic violence required to perpetuate those practices and to ensure the continued existence of related socio-ecological inequalities—inequalities that are further reinforced through overlapping legal, political, and spatial practices (Razack 2002; Gilmore 2021). These forms of institutional violence and discrimination are also integral to the logic and execution of racial capitalism. Recognizing this, significant EJ work has exposed one such form—environmental racism—and has developed effective political and legal strategies to challenge the institutions responsible for it. Environmental racism is understood as "any policy, practice, or directive that differentially affects or disadvantages (whether intended or unintended) individuals, groups, or communities based on race or color" and which is "reinforced by government, legal, economic, political, and military institutions" (Bullard 1994, p. 451). This definition of environmental racism underlines the fact that intentionality is not a decisive factor in establishing culpability for the violent effects of institutional operations. Identifying, and working to prevent, those pervasive effects has long been a central focus of EJ activism.

Ongoing practices of racial capitalism and environmental racism, fueled by growing amounts of energy, not only produce environmental inequalities but also (re)produce assumptions about societal organization, economics, politics, and nature. Environmental harms and forms of 'infrastructural violence' associated with energy infrastructures and capitalist production actively reinforce the racialized hierarchies and notions of difference that operate (and are imposed) across societies (Mbembe 2004; Rodgers and O'Neill 2012). This perspective is vital for understanding—and changing—contemporary energy relations. Human suffering and ecological degradation are not coincidental 'impacts' of globally dominant energy systems and energy practices. Socio-ecological inequalities are also effectively 'inputs' to most of those systems and practices (since their design and operation assume those inequalities to be inevitable or permissible). While the notion of *impacts* conveys a particular sense of temporality—attributing a neutrality to energy systems, as if society and nature could be reconfigured in ways that would enable them to better cope with the degradation being caused—a focus on systemic *inputs* highlights instead how the production of value and the categorization of difference are inter-related processes.

Combining these theoretical and empirical threads, many scholars conclude that environmental loss, disposability, and the racialized and unequal 'differentiation of human value' are inevitable within capitalist production (Melamed 2015, p. 77). This is not to attribute a monolithic, all-encompassing power or coherence to capitalism but instead to engage with capitalism as it emerges in patterns of human activity in diverse and contingent forms (Cross 2014, p. 32). The contribution of such EJ-oriented work to energy research is to maintain a critical focus on how energy practices and racial violence are simultaneously enmeshed within the processes, practices, and products that constitute racial capitalism. While the precise dynamics of these relationships vary across time and territories, there is a clear correspondence: the more that global societies depend on increasing energy use, the more we see worsening social and environmental effects of racialized systems of value production.

Energy as 'the Capacity to do Commodifiable Work'

Contemporary understandings of energy as industrialized work emerged from within the dominant worldview of anthropocentric white supremacy that fueled colonial acts of genocide and the Trans-Atlantic Slave Trade—a worldview which continues to create environmental injustice today by devaluing nature and non-white bodies (Lennon 2017). As noted above, critical perspectives on processes of industrialization highlight the importance of (re)historicizing the productive relations that constitute contemporary energy systems—specifically by analyzing the role of energy in the political, economic, and power relations that support the racialized production of differential value. EJ theory and praxis also pursue these analytical and political goals by re-examining specific relationships between energy and the *politics of work*.

Crucial here is disentangling energy in the abstract from how different forms of energy are experienced, located, and put to use—to recognize the term 'energy' as a social construction shaped within a specific historical context (e.g., Illich 2013). The standard definition of energy—'the capacity to do work'—is used in physics and thermodynamics to refer to a pervasive, universal force animating matter and motion. That empirical definition emerged in parallel with the socio-economic notion of a distinct 'energy sector' in the 1800s. The central importance of 'work' in both concepts reflects societal and ideological shifts that occurred throughout periods of industrialization—specifically the goal of formalizing, intensifying, and disciplining labor in order to maximize industrial production (Lohmann and Hildyard 2014; Daggett 2019). The notion of energy as mechanical work provides a useful term for referring to diverse types of fuel and power but, at the same time, carries with it a number of political implications. The energy-as-work concept dominates historical accounts of energy and development,[1] often with the effect of mischaracterizing difference and conflict within those processes.

How the relationships between energy and work are conceptualized and actualized materially affects who works, who gets paid for work, and how work is more broadly understood. Consistently driven to maximize production (and profit), energy-intensive industrialism has shown few signs of being focused on saving labor or eliminating work and slavery. Following Lohmann and Hildyard (2014, p. 44), it is arguably more accurate to say the industrial energy regime is a product of slavery not a replacement for it. Johnson (2016, 2019) makes a similar point, arguing that the broad historical term 'industrialization' itself reinforces a particular, generic view of work as a quantifiable product for sale and exchange, collapsing important ontological distinctions between human bodies and machines. When people are seen as mere agents of energy, the distinct—and markedly differential—experiences of actual people are lost.

The use of coal and steam technologies, building on the violent industrialization of human labor through slavery, effectively translated diverse forms of mechanical and chemical energy into flexible, transferable forms of production unbound by place or diurnal/seasonal cycles (Malm 2016). This had the effect of making labor more abstract, calculable, productive of surplus, and subject to commodification (Lohmann and Hildyard 2014). Energy-as-work created a corollary concept of work-as-energy where human effort was interchangeable with (or determined to be in competition with) mechanical power.

---

[1] For a detailed review of work on changing energy-labour relationships; connections between capital-labour regimes and social reproduction; and an account of an emergent 'energy labour' perspective in critical thought, see Bouzarovski (2022).

Distinctions between the movement of bodies and the power of machines were further blurred by the commodification and industrialization of labor. Silvia Federici argues that the productive forces of capitalist industrialization were predicated on the mechanization of workers' bodies and the transformation of women's bodies into machines to produce new workers, such that "the human body and not the steam engine, and not even the clock, was the first machine developed by capitalism" (Federici 2004, p. 146). Industrial histories show how the 'work' in the standard definition of energy is a dehumanized, generic form of work. The purposes to which that work is directed appear to be many, but those wide-ranging activities are reduced to fulfilling what is effectively a singular objective: the expansion of production for the maximization of profit. I propose that the common definition of energy is more accurately formulated as 'the capacity to do commodifiable work,' particularly when our goal is to better understand (and intervene in) the impacts of different energy systems upon people and places.

Some scholars have reworked the energy-as-work and work-as-energy corollary (and its problematic blurring of bodies and machines) as a tool of critique. For example, Fuller (1969) and McNeill (2000), among others, have conceptualized 'energy slaves' as the amount of work done by machines that has replaced human effort. Building on this idea, Mouhot argues contemporary economic and societal dependence on fossil fuels mirrors the dependence on bonded labor of openly slave-holding societies (Mouhot 2011, p. 343). It is an idea with a long history and has been used to quantify and compare the extent of energy-labor dependence across different epochs and locations. Even in the 1880s, French economist Emile Levasseur calculated the huge increase in the use of steam-powered machines in France in terms of the 98 million equivalent laborers whose efforts those machines had surpassed (Wrigley 1988). Focusing on oil in contemporary North America, Nikiforuk uses 12-hour days of human effort to calculate that the average citizen there depends on the work of 89 such 'energy slaves' since average annual per capita oil consumption is 23.6 barrels and each barrel is equivalent to 3.8 years of labor (Nikiforuk 2012). These formulations serve to re-emphasize the sheer scale of contemporary energy use and also offer another way to chart the ongoing growth in global energy demand.

Formulations of the 'energy slave,' however, carry political implications. Two, in particular, are problematic. First, discussing slavery in the abstract in these ways risks dehumanizing and commodifying slaves, their bodies, and their individual experiences—thus perpetuating the violence upon which systems of slavery are constructed (Pritchard 2015, p. 20). There have been moments when machines directly replaced slave labor, for example during labor shortages in ancient Rome; however, any energy-work equivalence misses how work-as-labor, both materially and analytically, is a far more complex socio-political phenomenon than the physical "expenditure of energy" (Debeir et al. 1991, p. 9). Second, rendering slavery as an abstract concept and discussing it without reference to race and racial violence draws attention away from how different people are exploited and minimizes the role of those who continue to profit from that exploitation—social inequalities that pivot around the production and use of energy. The term also obscures those inequalities. The "deracinated vocabulary" of the "energy slave" masks the persistence of exploitation and physical labor within current fossil fuel economies, while also hiding the socially specific experiences of suffering of workers from those whose everyday lives benefit from the emancipatory effects of mechanical energy use (Johnson 2016, pp. 973–974). Looking beyond the routine perspective of energy-as-work therefore involves scrutinizing the social and environmental effects of the mechanization and commodification of work.

Illich argued that "there can be no history of energy as a popular construct without a history of work, and vice versa" (Illich 2013, p. 114 [1983]). As previous sections have described, a focus on labor relations in energy histories highlights how dependent the growth of industrial energy regimes was, and is, on colonialism, racism, and the dual imposition-and-exploitation of differences in gender and class. The subjects of energy-as-commodifiable-work, therefore, vary across contexts (and according to the demands of particular industries).

The Industrial Revolution in England was initially heavily dependent on the labor of women and children, contrary to its common association with a male-dominated factory workforce (Foster and Clark 2018). An 1833 report on the workforce in eight key industries (including cotton and wool) across 271 factories in Great Britain showed that 56.8% of workers were women (Burnette 2008). The spinning jenny—a device designed to maximize the amount of cloth one worker could produce and which displaced more than nine in ten workers in some parts of England was originally intended for use by a young girl, built as it was around a horizontal wheel that made it uncomfortable for an adult worker to operate (Berg 1993, pp. 34; 40). Particularly when workers faced increasingly dangerous working conditions, maintaining such systems of age-based and gendered exploitation required differences and divisions within the working class to be aggravated—so that solidarity was discouraged and hierarchies built upon gender, race, and age could be imposed and perpetuated (Federici 2004, p. 63). Energy relations have therefore not only shaped the kinds of work available to people and the conditions that allow some people to profit by exploiting that work; energy relations also directly inform how labor (and its social power) is governed (and curtailed).

Analyzing energy-work relationships can also recast EJ questions concerning human-nature inequalities. Histories of energy—the study of how different energy flows and infrastructures take shape over time—are also environmental histories. Writing about the Columbia river in the Western US, White (1996) argues that a geography of energy is simultaneously a geography of labor, and energy here includes wind, waves, and river currents, as well as river dwellers such as the salmon whose energy exchange with humans and with the river itself have shaped this socioecological context. These are the reciprocal ties that connect humans with other biota in caring for one another and sustaining all life (Quaempts et al. 2018). Harnessing energy for the purposes of power generation thus leads not only to environmental disruption but also to struggles for social power and the ability to benefit from the labor of others (White 1996). As noted above, processes of industrialization have relied heavily on imposing and policing gendered and class-based hierarchies. Critical perspectives illustrate how the 'material particularities' of energy systems become socially and politically significant (Bridge et al. 2018). Centering EJ concerns within these histories shows how those same processes also impose and enforce power imbalances across differences in species.

Centering environmental justice in histories of energy has another analytical effect: shifting our analytical focus beyond 'energetic determinism' to see historical change as ecological change (Merchant 1990). Climate change as a result of carbon emissions in the atmosphere, the Anthropocene, the notion of "petroculture," oil spills, the toxic air that results from burning vast quantities of hydrocarbons—the type and scale of these energy-society relations are typically traced back to the processes of industrialization mentioned above, specifically the Industrial Revolution starting in Europe in the mid eighteenth century. However, there are histories of land use conflicts associated with growing energy needs that predate that particular period of economic expansion.

Medieval England had shipbuilding, soap, glass, iron, and copper-refining industries that all depended on forests for energy and, by the late sixteenth century, acute shortages of wood led to conflicts including uprisings by poor laborers and those dispossessed by land enclosures (Merchant 1990, p. 63). There were at least two key changes that occurred with the emerging capitalist economy that link directly to the ongoing climate crisis today: (i) dependence on organic or renewable energy sources (wood, water, wind, human and animal effort) was replaced with dependence on a nonrenewable energy source (coal); and (ii) the focus of production shifted toward processing inorganic materials (iron, copper, silver, gold, tin), all of which depended on extensive mining operations and further deforestation (and growing energy use) (Merchant 1990, p. 63). These apparent epochs, however, reflect not only shifts in energy demand and in the purposes for which energy (and work) were used; these shifts also depended on changing views of Nature, on what was considered to be the environment, and on the notion of resources being used to support the simultaneous exploitation of humans and Nature.

Energy transitions, then, are always bound up with changing social and environmental relationships of many different kinds, including labor relations. And those labor relations are not only linked to work and exploitation but also to how individuals and societies see and understand themselves. Building on White's research, LeMenager suggests that when we talk about energy, transitions, and the different dominant sources of fuel used within successive 'energy regimes,' we are concerned with how humans (and nonhumans) do work, avoid work, and experience that work being exploited or replaced by machines (LeMenager 2014). This focus keeps both structural and more (inter)personal dynamics within the analytical frame—an analysis of labor politics within energy relations that is also attuned to notions and experiences of subjectivity. For example, the expansion of fossil fuel consumption in the twentieth century, particularly in the US, saw the productive forces of capital extend into the reproductive spheres of everyday life, fostering an individual sense of liberation and power over space enacted through suburban home-ownership and auto-mobility (Huber 2013, p. xv). Individualized forms of mobility dominated twentieth century urban design, shaping the spaces where people live, work, and relax, and also shaping how those various spheres of human activity connect with each other. Energy systems continue to influence how labor relates to social organization, how people come to know nature, and how, through work, people relate to their environment (White 1996).

Those experiences vary, in turn, in relation to how people are differentially positioned geographically, culturally, and socially—accounts of subjectivity that critical feminist work has cast in new light.

While relationships between EJ movements and feminism are varied and unfixed, a number of insights from ecofeminism and feminist political ecology have been developed within EJ struggles and often outside of academia (Rocheleau et al. 1996; Martinez-Alier et al. 2014). By analyzing how the degradation of ecosystems and the subjugation of women are simultaneous and inseparable processes, materialist ecofeminism highlights how diverse bodies and beings are systematically made 'other' and devalued under patriarchal capitalism. Destructive relationships between human societies and the beyond-human world are predicated on the positioning of women (and others) at the boundaries of economic systems; the externalization and exploitation of women and of nature are materially linked (Mellor 2006, p. 140). Analyzing those links and the forces that mediate relationships between women, men, and nature, works in materialist ecofeminism critique essentialist views that categorize women's bodies as being 'closer' to Nature and instead refocus on labor, inequality, and colonialism (Mies and Shiva 1993; Sydee and Beder 2001). Gendered categorizations in patriarchal capitalism value men's productive labor as removed from nature (while women's reproductive labor remains "in" nature); materialist ecofeminism therefore locates the fundamental contradiction of capitalism not between capital and labor but instead "between production and reproduction" (Sydee and Beder 2001, p. 282). The material impacts of energy relations, then, are witnessed not only in the changes that energy systems inflict upon human and environmental health and wellbeing but also in the societal replication of highly gendered processes of valuing and devaluing.

In addition to highlighting gender disparities in access to energy and participation in energy-related decision making, critical ecofeminist approaches assess what kinds of productive ends energy is being directed towards, asking what relations of re/production are being supported (or undermined) by particular energy projects. This means bringing to the fore the often obscured political (and productive) relations that both constitute and are constituted by energy systems. Fraser (2014) notes that the devaluation of reproductive work co-occurs with the depoliticization of the environment (and, by extension, of energy systems) as part of efforts to hide or deny capitalism's reliance on unpaid work, the Earth's ecosystems, and political support for private enterprise (Bell et al. 2020, p. 4).

Feminist-Marxist approaches similarly scrutinize the naturalization of exclusionary labor relations under capitalism, analyzing the role of care and unpaid work in the production of value (Müller 2019). Whenever energy research replicates or relies on the straightforward theoretical equivalence of energy with work, it risks contributing to these dual, and interlocked, unjust processes: (i) denying and devaluing the economy's fundamental dependence on healthy, regenerating ecosystems, and (ii) denying and devaluing non-mechanical work.

Re-centering devalued forms of labor within energy analyses has the effect of challenging modern notions of relationality, subjectivity, and society (imagined as a collection of individuals, each bound to laws of economic necessity and connected to others primarily through acts of work)—and this has direct consequences for how environmentalism and environmental justice are understood (Young 2019). When analysts proposed post-materialism as an explanation for the rise of wilderness environmentalism from the 1970s onwards in high-consumption countries including the USA—the idea that people turned to 'post-materialist' pursuits such as enjoying time in a beautiful outdoor environment once their immediate material needs had been met within a consumer society—it was as if they believed "growing popular interest in nature was not so much a rejection of the modern world as a proper fulfilment of it" (Guha and Martinez-Alier 1997, p. xiv). Countering this idea, scholars in the field of ecological economics illustrated how post-materialism mischaracterized the material basis of environmentalism in high-consumption contexts and misunderstood the motivations for environmentalism in low-consumption contexts.

Post-materialist perspectives obscured from view both the vast flows of energy and materials that support 'environmental' pursuits in affluent societies as well as the legitimate concerns residents had regarding increasing environmental damage, risk, and pollution (Martinez-Alier 2002). Worse still, the implication of post-materialism was that poor and marginalized groups around the world were not active environmentalists—unable to prioritize care for the environment while struggling to meet subsistence needs—thus obscuring how the material use of "nature's life-support systems" often co-exists with localized practices of environmental care (Martinez-Alier 2002, p. 254). Against post-materialism, EJ work has documented countless cases of subsistence practices of environmental care, acknowledging and centering the many kinds of knowledge and expertise required. This work illustrates some of the different ways in which environmental action grows out of a recognition that the well-being of nonhuman entities and relations is necessary for human societies to survive and thrive. These are forms of environmentalism that emerge as an extension of working directly with, as in working within, the limits and regenerative cycles of the natural world.

Repoliticizing energy in line with critical perspectives on work and production also draws attention to questions of economic and ecological limits. The Marxist materialist critique of energy links the ecological impacts of high levels of energy use not to individual consumer habits, as is frequently argued in high consumption societies, but to the broader production, circulation, and accumulation of value under capitalism and the "material origins" of the commodity form—"what it takes to make a thing and what it takes to move it" (Bellamy and Diamanti 2018, pp. ix–x). This builds on Marx's concept of the 'metabolic rift' which refers to disruption of mutual relations between humans and the earth: a product of capitalism eroding the conditions for its future reproduction by exhausting or destroying natural resources and extracting nutrients from soils which it does not replace (Foster 2000; Nonini 2006). The changing social metabolism of industrial economies—the whole societal system of appropriating, transforming, and disposing of materials and energy—exacerbates environmental conflicts as new materials are sought from ever-expanding resource frontiers (Martinez-Alier et al. 2016). The rift was, and is, global: settlers robbed land, resources, and soil from colonies to fuel industrialization at home (Foster 2000). The history of energy is also the history of the rift.

Industrial production increased energy use in previously unimaginable ways. Using the solar energy of hundreds of millions of years ago stored as coal to power machines meant that production was no longer limited, as before, to annual solar flows (Marks 2002). In a letter to Marx, Engels described the 'working individual' as "not only a stabaliser of *present* but also, and to a far greater extent, a squanderer of *past*, solar heat" (Marx and Engels 1975, p. 411 [original emphasis], cited by: Foster 1999, p. 385). Echoing Lohmann and Hildyard's (2014) argument, cited above, that the industrial energy regime is more like a product of slavery than a replacement for it, Debeir et al. (1991) argue that the industrial use of steam engines was not intended to improve the conditions of labor for factory workers but rather to enable factory owners "to produce more, faster, and at lower costs" (Debeir et al. 1991, p. 7). The goal was simply to maximize the extraction of profit, no matter what the social cost. Energy relations were therefore central to Marx's vision of a future society. In this vision, 'associated producers' would collectively manage social-ecological metabolism in context-appropriate ways with 'the least expenditure of energy' (Foster 1999, p. 382, citing Capital vol. 3). Diverse contemporary movements—including EJ communities, Transition groups, and degrowth activists, as detailed below in Chap. 2—articulate a range of practices that similarly seek to end destructive patterns of ever-increasing resource and energy use. Echoing Marx, they recognize that energy relations can be redesigned to better fulfil a wide range of social needs, way beyond a singular focus on the profit motive.

Repoliticizing energy-work relations also acts to counter dominant perspectives whereby energy is naturalized, privatized, individualized, and therefore depoliticized—as happens with work when it becomes reified as a universal phenomenon removed from customary political contestation (Weeks 2011, p. 7; Daggett 2019, p. 198). The depoliticization of work not only limits how work might be reimagined to serve entirely different social, economic, and political ends, it also limits the ways in which new, more just, 'energy cultures' might be imagined (Daggett 2019, p. 197). As Malm (2016) underlines, the sphere of human praxis that puts the 'anthro' into 'anthropogenic climate change'—that directly links the violently unequal organization of human societies to planetary destruction—is labor (Malm 2016, p. 12). A critical perspective on energy-work relations therefore casts new light on how society intervenes in physical energy flows which then become embodied in institutions, infrastructures, and resources that are disputed by groups with distinct economic, cultural, and political projects. Doing so invites reassessment of the purposes of energy. Re-examining the relationships between energy use and work, building on EJ critiques, further opens up the potential for reconfiguring energy practices within prefigurative, emancipatory, and radical political projects.

Extractivism

Drawing variously on environmental history, ecological economics, critical social theory and feminist approaches, EJ also connects with a number of fields that scrutinize how energy practices relate to the extractive economies of modern industrialized society. Work in energy humanities includes analyses of how mass consumption, global mobility, and even many existing forms of the state and political organization are realized in relation to particular energy regimes (Szeman and Boyer 2017). Metabolic analyses, explored in more detail below (Chap. 2: Degrowth), show how growth in energy demand is in constant tension with environmental health and well-being. Political ecology scrutinizes the asymmetrical power relations that configure patterns of dominance and the exploitation of peoples, places, and ecosystems. Central to these analyses is scrutiny of how dispossession (of territories, sovereignties, resources, and rights) continues today through a variety of legal, economic, and military means, including the forceful expulsion of people from their lands and the privatization of common property (Harvey 2004). EJ movements are among the diverse protagonists at the forefront of resisting these overlapping forms of marginalization and discrimination—exposing and undoing practices of *extractivism* in multiple locations and across multiple scales.

The concept of extractivism draws together analytical critiques of how energy, economic growth, modernity, and global inequalities intersect—while also recognizing diverse forms of global resistance to the changes wrought by those processes and phenomena. Extractivism is "the appropriation of Nature to feed economic growth and the idea of development understood as an ongoing, linear process of material progress" (Gudynas 2013, p. 165). The origins of extractivism are those of the global economy: "extractivism is a mode of accumulation that started to be established on a massive scale 500 years ago. The world economy—the capitalist system—began to be structured with the conquest and colonization of the Americas, Africa and Asia" (Acosta 2013, p. 62). Extractivism is evident in the intensive exploitation of natural resources of all kinds—including fossil fuels, timber, rocks and minerals such as copper, gold, and lithium, all extracted for sale on global markets—that continues to drive development policy in countries across South America and elsewhere (Gudynas 2013). Critiques of extractivism show how the construction and growth of certain energy systems consolidate global patterns of unequal development.

While there are multiple overlapping definitions of extractivism, what they share is an understanding of extractivism as fundamental to the operation of capitalist and imperialist interests. Within that overall perspective, the emphasis may be on extractivism as an "ideological mindset" fixated on maximizing profit by maximizing environmental removal of material (Jalbert et al. 2017; Willow 2018; Yazzie 2018); on the 'overexploitation' of natural resources destined primarily to global export markets (Hamouchene 2019, p. 5); or on how successive waves of extractivism have dominated development thinking through ever-increasing use of water, energy, and other resources (Veltmeyer and Petras 2014; Svampa

2019). By definition, extractivism has no long-term future. It refers to economic and industrial activities that will ultimately exhaust the material basis of those activities. Extractivism is a pursuit whose "social and environmental costs are not included in the prices of products" since those costs are "externalized and carried by a society without democratic rights in a transnational entrepreneurial world" (Acosta 2013, p. 62). Extractivism dominates through the power of certain countries and corporations; it is both a major cause and consequence of ever-increasing global energy demand; and it is centered on coupling the privatization of profit to the socialization of harm.

Extractivism reflects particular ways of seeing and relating to the world, regarding Nature as a mere source of inputs for processes of accumulation. Analyzing extractivist policies and practices is a core component of critical energy research—another way of scrutinizing how energy relates to dominant systems of production and to the uneven distribution of profits and costs. This applies to emergent energy systems as much as it does to legacy energy operations. As Willow (2018) makes very clear, in part through case studies of large-scale hydroelectric generation, even though energy *extraction* might be more commonly associated with oil derricks and coal mines, *extractivism* is replicated across both non-renewable and renewable energy sources whenever projects are designed according to the dual goals of maximizing how much energy and how much private profit is generated (with minimal regard for negative social and environmental impacts). Critiques of extractivism are therefore important for identifying energy initiatives—including those presented as 'green capitalism'—which are likely to deepen rather than alleviate already-existing social and environmental injustices.

Hamza Hamouchene analyzes the Ouarzazate (Noor) solar plant in Morocco: a concentrated solar power project that acquired 3000 hectares of communally owned land to produce energy for both domestic and European export markets. His analysis centers on questions that challenge any proposed or actual energy infrastructure: Who owns what? Who does what? Who gets what? Who wins and who loses? Whose good is being served? (Hamouchene 2016, p. n). The Ouarzazate (Noor) project is emblematic of 'green capitalism' in that it proposes an environmental or 'climate friendly' initiative but does nothing to challenge the entrenched power of certain countries and corporations—thus deepening global inequalities in both economic power and in energy use—and sees economic benefits of the initiatives accrue to a relatively small number of people (Newell and Paterson 2010; Newell 2021). Based on a core idea in green capitalism—that "contributions to the improvement of the global environment should be sought where this is cheapest"—the project thus depends upon the "conventional economic notion of differential opportunity costs" which reinforce "inequalities between poorer and wealthier landholders, between urban and rural areas, and between the global South and North" (Hamouchene 2016, p. n). Diverse EJ resistance movements continue to challenge green capitalist initiatives and the threats they present to socioecological wellbeing (Santiago et al. 2017; Avila-Calero and Sorman 2018; Mendoza et al. 2021). These movements reject green capitalism as part of the 'strategic logic' of global economic growth and therefore also reject extractivist energy initiatives that replicate grossly unequal systems of decision-making and financing.

The effects of extractivism are witnessed globally. The Ouarzazate (Noor) solar plant is just one case where the unequal effects of extractivist policies and the privatization of natural resources—including communally held territories—have been exacerbated by processes of accumulation by dispossession across Northern African countries (Hamouchene 2019). In Mexico, Indigenous groups fight against the enclosure of communal lands and the private appropriation of benefits that accompany large-scale wind generation projects (Avila-Calero 2017). Rural coastal communities in Puerto Rico struggle against fossil fuel economies and new forms of solar-driven energy coloniality, while also mobilizing to transform how different forms of power are used, and by who (de Onís 2021). While experiences of extractivist energy regimes are distinct in each context, related patterns of exploitation, inequality, and irreversible ecological change emerge at national, regional, and global scales. Powell (2018) argues that deciphering these 500-year-old patterns—and building networks of resistance to them—involves analyzing energy and extractivism in relation to race, power, capitalism, and climate change, always recognizing that these overlapping processes are shaped by practices and ideologies of settler colonialism.

SETTLER COLONIALISM AND SOVEREIGNTY

Settler colonialism is an ongoing structure of violence and exploitation that calls for the annihilation of Indigenous peoples and seizure of their territories (Estes 2019; Kojola and Pellow 2021). It is a form of ecological domination that violently disrupts relationships between humans and the world around them, and which undermines the resilience and rights of Indigenous peoples as self-determining collectives (Whyte 2018, p. 125). Settler colonialism is integral to capitalism and industrialization, affecting both those who resist (and suffer from) and those who perpetrate (or benefit from) these processes of domination—but diverse Indigenous peoples around the world are specifically targeted. As Kim Tallbear underlines, settler colonialism works to violently remove Indigenous peoples from their lands *and* to sever Indigenous peoples' relations with their lands (TallBear 2019). Racial capitalism and settler colonialism are closely linked as structures of oppression, and many of those links are to be found in the exploitation and extraction of resources for industrialized energy use. However, while racialization and colonization are both global systems that "secure white dominance through time, property, and notions of self," important differences remain—particularly in the US context: "American Indian national assertions of sovereignty, self-determination, and land rights disappear into U.S. territoriality [if] indigenous identity becomes a racial identity and citizens of colonized indigenous nations become internal ethnic minorities within the colonizing nation-state" (Byrd 2011, p. xxiv).

Analysis of settler colonialism is fundamental to understanding the production of environmental injustices, but the experiences of a diverse range of racial, political, and marginalized groups cannot be collapsed into a singular EJ framework (Gilio-Whitaker 2019; Kojola and Pellow 2021). The specificities of environmental (in)justice in Indigenous contexts are made distinct both through unique historical, political, and legal processes and because of forms of violence that are experienced specifically by Indigenous peoples (Grijalva 2008; Ranco et al. 2011; Jarratt-Snider and Nielsen 2020; McGregor et al. 2020). Understanding contested sovereignty is central to understanding these dynamics. In the US EJ Movement, actions affirming sovereignty and self-determination led by Native groups were acknowledged as distinct political strategies from those led by other EJ groups. This is reflected within the Principles of Environmental Justice from the first National People of Color Environmental Leadership Summit in 1991—but that distinction still has to be defended against mainstream environmentalist views that obscure Native sovereignty (POCELS 1991; Ishiyama and TallBear 2001). Recognizing differences and specificities across EJ and Indigenous struggles is integral to critical energy research.

Settler colonial theory underlines how the states that exist today as a result of colonization across the Americas (and around the world) are fundamentally structured upon—and thus continue to reproduce—violent and exploitative racial and gender logics (Speed 2017, p. 789). Energy development frequently brings these intersecting issues to the fore and is itself a terrain of political struggle. For example in the Navajo Nation, a region rich in energy minerals, resource politics combine negotiations over indigeneity, sovereignty, and the role of social movements in collective governance—all within a context that is overdetermined by a colonial configuration of relations of inequality (Powell 2018; Yazzie 2018). Even with formal sovereignty, native nations cannot initiate completely independent energy projects: the term 'resource colonialism' highlights how realizing Navajo sovereignty within energy initiatives, for example, is largely dependent on enacting (and interacting with) a regional (and national) resource economy of extraction and exploitation (Curley 2018; Yazzie 2018, p. 29). Simpson (2014) describes this as 'nested sovereignty': the political positioning of Indigenous nations and communities whereby Indigenous sovereignty is always contained and curtailed by settler sovereignty (Powell 2018, p. 5). Engaging with such diverse and contested meanings, experiences, and practices of sovereignty has important implications for future energy research—specifically by reframing (or rejecting) the concepts of recognition and participation so that they are better aligned with actual demands made by marginalized communities.

While 'justice as recognition' is a core component of much EJ and energy justice literature, it is frequently interpreted as merely involving increased attention to historically marginalized groups. By contrast, engaging with sovereignty struggles shows how recognition justice is more meaningful (i) when its goal is not identifying subjects or stakeholders but is instead empowering new agents and leaders, or (ii) when recognition justice necessitates a direct challenge to the singular state-oriented power relations that are embedded in energy systems (Avila 2018). In many contexts, any sense of justice is impossible using existing recognition paradigms: US state agencies formally 'recognizing' Indigenous sovereignty, for example, but doing so within existing legal-juridical frameworks, does little to support collective efforts to confront and reverse environmental injustices (Whyte 2017b). Further steps are also required (building on Simpson's critique of how 'nested' sovereignty curtails collective actions), such as recognizing internal diversity and power structures within affected communities (Ishiyama and TallBear 2001) and acknowledging how sovereignty is differentially lived and defined by distinct nations, tribes, and peoples—often as a process that is embodied, relational, grounded, and in constant motion (Byrd 2011; Powell 2015). Sovereignty seen like this, as ongoing practices and processes, exceeds the constraints of settler-state legal systems.

Similarly, far from creating a context of 'procedural justice,' Indigenous and marginalized groups have found the notion of increased 'participation' in state decision-making processes has only served to reinforce hierarchies and inequalities when carried out according to terms that have been defined by state agencies and/or enforced by discriminatory judicial systems (Goldtooth 1995; Weaver 1996; Ranco 2008). Many sovereignty struggles therefore embody ways of relating, interacting, and generating forms of social organization that explicitly challenge the assumption that the state is the center of political life (Goodyear-Ka'ōpua 2011). De-centering the state, particularly within struggles for sovereignty and decolonization, is a response to how 'recognition' can have harmful or contradictory consequences for those pursuing justice claims.[2]

Audra Simpson (2014) specifically proposes a politics of refusal as a political alternative to 'recognition.' Analyzing American Indian relations to the settler state and drawing on struggles in the Haudenosaunee political context, refusal in this sense is a political and ethical position which demands that nationhood and sovereignty are both acknowledged and upheld—thus questioning the legitimacy of agencies, individuals, and institutions who typically assume the position of 'recognizing' others: "What is their authority to do so? Where does it come from? Who are they to do so?" (Simpson 2014, p. 11). In what is today called Canada, Native rights activists refuse the neocolonial petro-state's claims of legitimacy since those claims are based on ongoing acts of violence, the enclosure of the commons, and the deliberate dissolution of collective rights (Mookerjea

---

[2] It is important to note that formal state-based recognition can still, in some contexts, be a vital part of collective efforts to defend Indigenous lives, rights, and livelihoods. Even if such recognition is compromised by ongoing discrimination within existing legal-juridical systems, it can be a necessary step toward claiming and defending other (previously rejected or violated) constitutional rights or rights to land and collective self-governance (as has occurred, for example, in highland Ecuador, see Partridge 2017b, 2018). In the context of Louisiana and the fight for recognition pursued by the Grand Caillou/Dulac, Isle de Jean Charles, and Bayou Lafourche Tribes, seeking federal recognition—while seen as being both a step toward protecting rights and also potentially, problematically, re-inscribing US colonial authority—can benefit justice struggles by supporting claims for compensation (as groups who have suffered the environmental and health impacts of corporate malfeasance, violence, and negligence) and also as a form of resistance against the institutional erasure of Native histories (Maldonado 2019).

2017). Acts of refusal are therefore also acts of affirmation, 'amplifying and centering' Indigenous lives and livelihoods (Simpson 2021). Refusal is distinct from resistance: it positively asserts Indigenous nationhood and peoplehood while negating the assumption that authority is a given (Grande 2018). While Simpson's work refers to the specific contexts and struggles of Indigenous political actions, a review of her book *Mohawk Interruptus* offers a summary of how refusal might also enable others to reformulate ideas about *resistance* and therefore also about *justice*:

> Capitalist settler states prefer resistance, which can be negotiated or recognized... Refusal, on the other hand, interrupts the smooth operation of power, denying presumed authority and remaking ignored narratives... Where resistance looks for lacuna and interruptions in the constancy of power, refusal denies its very legitimacy... Where the modern nation-state insists on its own totalizing reality, those who refuse deny its authority and its domination. Where the state claims a monopoly on violence, those who refuse deny its universality and capacity. (Ferguson 2015, p. n)

There are many diverse Indigenous movements for sovereignty and plurinationality that further reveal how contemporary formations of the nationstate routinely carry out exclusionary and destructive practices. These are decolonial and anticolonial movements that prioritize plurinationality and diversity together with "the sovereignty of indigenous people over their own territories, autonomy of peoples, communities and movements, judicial pluralism, the rejection of the developmental state and extractivism, as well as the recognition of the rights of Mother Earth" (Lander 2013, p. 90). In Ecuador, Indigenous nationalities successfully campaigned for the national constitution to define the country as plurinational—something that supporters maintain is not only about the protection of the autonomy, self-determination, and territorial rights of distinct Indigenous nationalities but also a national demand to deepen democracy for all peoples within the country (Acosta and Martínez 2009). Yet, despite these legal gains, Indigenous people (in particular) and EJ movements are among those groups who find their rights continually under threat—including rights that are legally acknowledged—often as a direct result of state actions.

A goal, therefore, of critical energy research is not to reform or further refine the specification of what more effective (or 'just') kinds of recognition might involve. Instead, the goal is to reconsider *whose* power is placed at the center of struggles for justice. In many cases, this represents an inversion of the 'recognition' paradigm and its priority: in place of the dominant practice of considering which groups are currently under-recognized by powerful state and corporate institutions (and then petitioning those actors to modify their behavior and decision-making processes), the inverse approach instead reconsiders whether the power held (and authority claimed) by those powerful agents are to be 'recognized' (or considered likely to play a constructive role in realizing justice objectives) by those making justice claims and demands. Actively pursuing this inverse approach in energy research could involve respecting spaces in which political *refusal* is the basis of struggles for justice—and (re)formulating justice claims, when appropriate, in light of those political actions. This is one way among others—including movements for autonomous and prefigurative politics—to address the role of energy systems in shaping how discrimination and power are differentially experienced or resisted across societies.

Autonomy, Prefiguration, and the State

A politics of refusal is only one formulation among many where the role of the state is reconsidered and reframed. EJ work draws on critical theory to scrutinize how the state itself is a site of racial conflict: where the social structures and representations that underpin racial categories are reinforced (Omi and Winant 1994; Goldberg 2001); where institutions delimit how environmental injustices are identified and addressed (Kurtz 2010); and where, far from being a "neutral force," state agencies routinely sanction various forms of environmental degradation and racial violence (Pulido 2017, p. 524). Pellow highlights the view that nation states are "both inherently racially exclusionary and ecologically unsustainable" since their purpose is to "exert control over populations, ecosystems, and territory, among other things" (Pellow 2018, p. 33). EJ movements around the world have regularly come up against the limits of these exclusionary practices. International governing bodies and NGOs also tend to be locked into the procedural norms of state administrations, limiting any ability to operate beyond the entrenched, discriminatory practices of national legal and political mechanisms (McGregor et al. 2020). And the boundaries that define nation-states frequently bifurcate communities and complex bioregions—such as the world's largest mangrove forest across the Sundarbans of India and Bangladesh—fragmenting collective governance strategies and creating further ecological harm (Kothari 2014). Indigenous and critical EJ perspectives on the state accordingly call into question the potential for systemic injustices to be resolved within existing state, legal, and economic frameworks.

Seeking to minimize dependence on limiting and discriminatory state logics (and questioning the reliability of the state for delivering justice), a wide array of alternative practices has emerged across the global EJ movement and in allied activist spaces. When relations with the state involve conditions of conflict, compromise, exclusion, and exploitation, such alternatives become necessary. These diverse practices explore autonomy, communality, relationality, anarchistic and horizontal political strategies and, when engagement with the state is unavoidable, strategies for what Escobar refers to as 'making counterhegemonic use of hegemonic tools' (Santos 2007; Escobar 2020). Critical EJ engagements with the state similarly tend toward the strategic: building movements that consistently challenge state power while acknowledging that sometimes state agencies can be pushed to support justice goals; and always working collaboratively to support and develop practices of direct democracy (Pellow 2018). The Critical EJ focus on deepening direct democracy is therefore not necessarily aiming to "abolish the state" but rather seeking to abolish the "socio-ecologically violent, hierarchical relationships that tend to support state institutions and flow from them" (Pellow 2018, p. 24). This connects to a fundamental concern for a critical energy research agenda: only when the currently dominant forms of *enacting* power in society *and* the currently dominant forms of *addressing* power asymmetries are radically reassembled—by centering the lives, livelihoods, power, and interests of people currently excluded from the decision-making processes that most affect their lived experiences—only then can any kind of meaningful 'energy justice' be talked about, let alone achieved.

Although critics question whether all citizens want to be consistently involved in collective governance (and also whether direct democracy has the same capacity as existing state structures for coordinating planning and resources on very large scales), direct democratic processes provide a flexible model for rethinking how people interact and cooperate—inviting reconfigurations of how power is established and exercised (Zografos 2019). While collective self-rule and shared responsibility are key characteristics of direct democracy, a broad palette of further practices can help foster genuine participation (rather than mere consultation); these include consensus decision-making, community budgeting, skill sharing, and popular education (Chatterton and Cutler 2008). Similar approaches include practices of 'counterpower'—drawing on direct democracy traditions as well as consensus decision-making and mediation—that create spaces for the public negotiation, control, and transformation of social systems toward shared forms of value (such as conviviality or prosperity) (Graeber 2004).

Engaging with these principles, the concept of 'energy democracy' has shaped a wide range of community responses to new energy initiatives, including specific proposals that move away from standard state- and corporate-run models in favor of cooperative ownership, community-led municipalism, and alternatives to growth-oriented development. Diverse, emergent social movements continue to mobilize around energy democracy as a way to re-imagine and re-design energy systems such that they are more attentive to issues of justice, popular participation, and social power (Endres and Johnson 2022). Energy democracy movements in New York City and elsewhere in the US, many of them engaged with the Solarize model of expanding residential solar generation, draw together a number of critical threads: an analysis of racial capitalism; pushing for an equitable transformation of energy infrastructures in low-income communities of color; generating well-paid and secure renewable energy jobs in those communities; and shifting energy governance from investor-owned companies to publicly-controlled organizations (Fairchild and Weinrub 2017; Lennon 2021, p. 4). Parallel movements in Louisiana make the case for community solar initiatives as a form of 'energy reparations' by relocating control of energy systems, and the multiple forms of power associated with that control, within low-income and Black communities—as both a form of reparation and as a forward-looking strategy for dismantling the processes of slavery, patriarchy, imperialism, and genocide that fuel a status quo defined by 'petro-racial capitalism' (Luke and Heynen 2020).

Further definitions—and practices—of energy democracy vary across contexts in ways that reflect how democracy itself is variously enacted and contested, denied or defended. The extent to which energy democracy initiatives depend on state institutions similarly varies across contexts. A common ethos of energy democracy movements, though, is the repositioning of all those associated with energy systems—consumers, laborers and producers, decision-makers, underserved communities—and finding ways to revise (or strengthen) the relationships that connect those different groups.

Other core principles of energy democracy include greater transparency and public participation—but participation that goes beyond merely inviting community members into extant decision-making processes in order to also open the possibility of restructuring those processes such that community members can take on actual roles of leadership. This can mean that communities play an active role in deciding how the energy they consume is generated and also in deciding what types of distributing organization or institution are responsible for making that energy accessible (Baker 2016). By restructuring or removing particular institutions, these renewed approaches to energy democracy aim to both counteract the concentration of wealth and power associated with multinational fossil fuel companies (system change and localization) and also promote diversity and inclusion within decision-making processes (system design and participation) (Stephens 2019; Partridge forthcoming). Communities may therefore seek greater freedom from the confines of state structures, either as a way to protect against injustices (Scott 2010) or as a means to enact desired social change on terms that participants can set themselves (Pellow 2018). These characteristics of energy democracy are also characteristics of 'prefigurative politics,' where the interpersonal relationships, division of labor, and decision-making processes involved in the pursuit of political objectives embody (and anticipate or 'prefigure') a group's desired outcomes (Polletta 2002). Seeking justice through greater autonomy from the state thus becomes one way of refusing to reproduce or rely upon relations of violence, hierarchy, and exclusion that are themselves sources of injustice (Yuen 2001).

Movements engaged in prefigurative politics may enjoy only partial degrees of autonomy yet still exert an influence on broader struggles for justice. This happens, for example, when apparently localized struggles initiate political changes both within and beyond their contextual legal structures. Hydrocarbon-based economies are those in which privatized, market-based and extractivist initiatives dominate (Boykoff et al. 2009), typically to a point where alternative forms of economic and social life are denigrated, silenced, or merely painted as unrealistic: prefigurative justice movements are thus forced to act not *outside* but *despite* the political-economic systems they oppose (Partridge 2017a, p. 200). By creating and holding open prefigurative political spaces, justice movements can *precede* institutional change by illustrating and enacting the socio-environmental relations that reconfigure, redistribute, and re-orient power toward more just and equitable ends (Kenrick 2013). Questions of justice are thus not limited to better policy design, but rather involve actively supporting specific acts of resistance and pathways to change—such as bottom-up processes of organization and design that can address social, economic, and energy-related inequalities and vulnerabilities (Partridge 2017a). Prefigurative political action not only shows what might be possible in a hypothetical future *beyond* current political-economic constraints; such acts also illustrate how to pursue and embody resistance *against* those constraints and forms of oppression.

In India, movements for Decentralized Renewable Energy (DRE) have called for 'responsible' energy production governed by community-level systems of control and management (Hande et al. 2017). The governance of energy infrastructures presents challenges for direct democracy since—like large watersheds or cross-border forests—they tend to be large, complex entities that span multiple territories and jurisdictions, meaning that not all decisions can be made in face-to-face assemblies. Movements for DRE echo the actions of many global movements in anticipating these challenges and devising modes of organizing that are designed to counteract them. For example in such cases, where delegation is required, accountability can be maintained and the concentration of power avoided by implementing a system of rotation between different delegates—just as corrective pathways can be designed to counteract other challenges to direct democracy, such as persistent forms of gender, caste, or class discrimination within particular collectivities (Kothari 2019). Of course, like any collective endeavor in any global context, grassroots initiatives for direct democracy and energy democracy face the challenge of ensuring that they do not replicate, on a smaller scale, dominant forms of discrimination that exist throughout society as a whole. For this reason, as is the case with DRE, movements will often make this challenge an explicit focus of their work—hence the collective design of corrective pathways and processes.

The importance of such energy-focused movements—pursuing prefigurative politics or seeking greater degrees of political autonomy—is reflected not only in the immediate, local, and material changes that they bring about. By enacting their desired alternatives to environmental harm and exploitation, such movements are also important for how they expose the assumed *inevitability* of extant, dominant ways of organizing social, economic, and productive life across society(s). The heavily subsidized corporate extraction and combustion of fossil fuels—the hydrocarbon economy—dominates global energy and economic systems to the extent that these practices have become naturalized and are assumed to be inevitable (Jalbert et al. 2017; Partridge 2017a). A consequence of this is that *the future* has also been colonized—by singular visions of a market-based economy that continues to expand by consuming growing amounts of energy and extending the power and influence of transnational corporations (Goodyear-Ka'ōpua and Baker 2012). Operating according to different principles—principles that are generated collectively through difficult processes of struggle and negotiation, and which reject those colonized futures—energy-focused and EJ movements instead reaffirm the many possibilities that are obscured when the ongoing dominance of the hydrocarbon economy is assumed to be inevitable.

The work of these movements reaffirms that other worlds are possible. In part, by building energy systems that are designed to reduce or remove environmental injustices. In part, by expanding on the notion of energy transition to see it as a project that is more radical, systemic, and politically engaged (Healy and Barry 2017, p. 452). As explored in the next section—Chap. 2: Transitions beyond crisis—the process of formulating such (engaged, politicized) transition projects draws heavily from the already-underway efforts of diverse EJ and Indigenous movements for transformative social change.

References

Acosta, A. 2013. Extractivism and Neoextractivism: Two Sides of the Same Curse. In *Beyond Development: Alternative Visions from Latin America*, ed. M. Lang and D. Mokrani, 61–86. Quito: Transnational Institute.

Acosta, A., and E. Martínez, eds. 2009. *Plurinacionalidad: Democracia en la diversidad*. Quito: Abya-Yala.

Appel, H.C. 2012. Walls and White Elephants: Oil Extraction, Responsibility, and Infrastructural Violence in Equatorial Guinea. *Ethnography* 13 (4): 439–465.

Avila, S. 2018. Environmental Justice and the Expanding Geography of Wind Power Conflicts. *Sustainability Science* 13 (3): 599–616.

Avila-Calero, S. 2017. Contesting Energy Transitions: Wind Power and Conflicts in the Isthmus of Tehuantepec. *Journal of Political Ecology* 24: 992–1012.

Avila-Calero, S., and A. Sorman. 2018. Transición energética (energías renovables). In *Decrecimiento: Vocabulario para una nueva era*, ed. G. D'Alisa, F. Demaria, and G. Kallis, 360–364. Barcelona: Icaria Editorial.

Baker, S.H. 2016. Mexican Energy Reform, Climate Change, and Energy Justice in Indigenous Communities. *Natural Resources Journal* 56 (2). 369–390.

————. 2021. *Revolutionary Power: An Activist's Guide to the Energy Transition*. Washington: Island Press.

Bell, S.E., C. Daggett, and C. Labuski. 2020. Toward Feminist Energy Systems: Why Adding Women and Solar Panels is Not Enough. *Energy Research & Social Science* 68: 101557.

Bellamy, B., and J. Diamanti, eds. 2018. *Materialism and the Critique of Energy*. Chicago: MCM'.

Berg, M. 1993. What Difference Did Women's Work Make to the Industrial Revolution? *History Workshop Journal* 35 (1): 22–44.

Bledsoe, A., and Wright, W. J. 2019. The anti-Blackness of global capital. *Environment and Planning D: Society and Space* 37 (1): 8–26.

Bouzarovski, S. 2022. Energy and Labour: Thinking across the Continuum. *Progress in Human Geography* 46 (3): 753–774.

Boykoff, M.T., A. Bumpus, D. Liverman, and S. Randalls. 2009. Theorizing the Carbon Economy: Introduction to the Special Issue. *Environment and Planning A: Economy and Space* 41 (10): 2299–2304.

Bridge, G., S. Barr, S. Bouzarovski, M. Bradshaw, E. Brown, H. Bulkeley, and G. Walker. 2018. *Energy and Society: A Critical Perspective*. London: Routledge.

Building Bridges Collective. 2010. *Space for Movement? Reflections from Bolivia on Climate Justice, Social Movements and the State*. Leeds: Footprint Workers Co-op.

Bullard, R.D. 1990. *Dumping in Dixie: Race, Class and Environmental Quality*. Boulder: Westview Press.

————. 1994. The Legacy of American Apartheid and Environmental Racism. *Journal of Civil Rights and Economic Development* 9 (2): 445–474.

Burnette, J. 2008. Women Workers in the British Industrial Revolution. *EH.Net Encyclopedia*, 26 March.

Byrd, J.A. 2011. *The Transit of Empire: Indigenous Critiques of Colonialism*. Minneapolis: University of Minnesota Press.

Carrington, S.H.H. 2003. Capitalism & Slavery and Caribbean Historiography: An Evaluation. *The Journal of African American History* 88 (3): 304–312.

Chatterton, P., and A. Cutler. 2008. *The Rocky Road to a Real Transition: The Transition Towns Movement and What it Means for Social Change*. Leeds: The Trapese Collective.

Cole, L.W., and S.R. Foster. 2001. *From the Ground Up: Environmental Racism and the Rise of the Environmental Justice Movement*. New York: NYU Press.

Cross, J. 2014. *Dream Zones: Anticipating Capitalism and Development in India*. London: Pluto Press.

Curley, A. 2018. A Failed Green Future: Navajo Green Jobs and Energy "Transition" in the Navajo Nation. *Geoforum* 88: 57–65.

Daggett, C.N. 2019. *The Birth of Energy: Fossil Fuels, Thermodynamics, and the Politics of Work*. Durham: Duke University Press.

Debeir, J.-C., J.P. Deléage, and D. Hémery. 1991. *In the Servitude of Power: Energy and Civilization through the Ages*. London: Zed Books.

Del Bene, D., A. Scheidel, and L. Temper. 2018. More Dams, More Violence? A Global Analysis on Resistances and Repression around Conflictive Dams through Co-produced Knowledge. *Sustainability Science* 13 (3): 617–633.

EJNA. 2009. *Energy Justice in Native America: A Policy Paper for Consideration by the Obama Administration and the 111th Congress.* Honor the Earth; Intertribal Council On Utility Policy; International Indian Treaty Council; Indigenous Environmental Network.

Endres, D., and T.N. Johnson. 2022. Energy Democracy at the Scale of Indigenous Governance: Indigenous Native American Struggles for Democracy, Justice, and Decolonization. In *Routledge Handbook of Energy Democracy*, ed. A.M. Feldpausch-Parker, D. Endres, T.R. Peterson, and S.L. Gomez, 51–65. London: Routledge.

Escobar, A. 2020. *Pluriversal Politics: The Real and the Possible.* Durham: Duke University Press.

Estes, N. 2019. *Our History Is the Future: Standing Rock versus the Dakota Access Pipeline, and the Long Tradition of Indigenous Resistance.* London: Verso.

Fairchild, D., and A. Weinrub, eds. 2017. *Energy Democracy: Advancing Equity in Clean Energy Solutions.* Washington: Island Press.

Federici, S. 2004. *Caliban and the Witch: Women, the Body, and Primitive Accumulation.* New York: Autonomedia.

Ferguson, J. 1990. *The Anti-politics Machine: 'Development', Depoliticization and Bureaucratic Power in Lesotho.* Minneapolis: University of Minnesota Press.

———. 1999. *Expectations of Modernity: Myths and Meanings of Urban Life on the Zambian Copperbelt.* Berkeley: University of California Press.

Ferguson, K. 2015. Refusing Settler Colonialism: Simpson's Mohawk Interruptus. *Theory & Event* 18 (4).

Fiori, N. 2020. Plantation Energy: From Slave Labor to Machine Discipline. *American Quarterly* 72 (3): 559–579.

Foster, J., and B. Clark. 2018. Women, Nature, and Capital in the Industrial Revolution. *Monthly Review* 69 (8).

Foster, J.B. 1999. Marx's Theory of Metabolic Rift: Classical Foundations for Environmental Sociology. *American Journal of Sociology* 105 (2): 366–405.

———. 2000. *Marx's Ecology: Materialism and Nature.* New York: Monthly Review Press.

Fraser, N. 2014. Behind Marx's Hidden Abode: For an Expanded Conception of Capitalism. *New Left Review* 86: 55–72.

Fuller, R. 1969. *Utopia or Oblivion: The Prospects for Humanity.* London: Allen Lane.

Gilio-Whitaker, D. 2019. *As Long as Grass Grows: The Indigenous Fight for Environmental Justice, from Colonization to Standing Rock.* Boston: Beacon Press.

Gilmore, R.W. 2021. *Change Everything: Racial Capitalism and the Case for Abolition*. Chicago: Haymarket Books.

Goldberg, D.T. 2001. *The Racial State*. Malden: Wiley-Blackwell.

Goldtooth, T. 1995. Indigenous Nations: Summary of Sovereignty and Its Implications for Environmental Protection. In *Environmental Justice: Issues, Policies, and Solutions*, ed. B. Bryant, 138–148. Washington: Island Press.

Goodyear-Ka'ōpua, N. 2011. Kuleana lāhui: Collective Responsibility for Hawaiian Nationhood in Activists' Praxis. *Affinities: A Journal of Radical Theory, Culture, and Action* 5: 130–163.

Goodyear-Ka'ōpua, N., and M.T. Baker. 2012. The Great Shift: Moving Beyond a Fossil Fuel–Based Economy. *Hūlili: Multidisciplinary Research on Hawaiian Well-Being* 8: 133–166.

Graeber, D. 2004. *Fragments of an Anarchist Anthropology*. Chicago: Prickly Paradigm Press.

Grande, S. 2018. Refusing the University. In *Toward What Justice?: Describing Diverse Dreams of Justice in Education*, ed. E. Tuck and K.W. Yang, 47–65. New York: Routledge.

Grijalva, J.M. 2008. *Closing the Circle: Environmental Justice in Indian Country*. Durham: Carolina Academic Press.

Gudynas, E. 2013. Transitions to Post-extractivism: Directions, Options, Areas of Action. In *Beyond Development: Alternative Visions from Latin America*, ed. M. Lang and D. Mokrani, 165–188. Quito: Transnational Institute.

Guha, R., and J. Martinez-Alier. 1997. *Varieties of Environmentalism Essays North and South*. London: Earthscan.

Hamouchene, H. 2016. The Ouarzazate Solar Plant in Morocco: Triumphal 'Green' Capitalism and the Privatization of Nature. *Jadaliyya*, 23 March.

———. 2019. *Extractivism and Resistance in North Africa*. London: Transnational Institute.

Hande, H., V. Shastry, and R. Misra. 2017. Energy Futures in India. In *Alternative Futures: India Unshackled*, ed. A. Kothari and K. Joy. Delhi: AuthorsUpFront.

Harvey, D. 2004. The 'New' Imperialism: Accumulation by Dispossession. *Socialist Register* 40: 63–87.

Healy, N., and J. Barry. 2017. Politicizing Energy Justice and Energy System Transitions: Fossil Fuel Divestment and a "Just Transition". *Energy Policy* 108: 451–459.

Healy, N., J.C. Stephens, and S.A. Malin. 2019. Embodied Energy Injustices: Unveiling and Politicizing the Transboundary Harms of Fossil Fuel Extractivism and Fossil Fuel Supply Chains. *Energy Research & Social Science* 48: 219–234.

Huber, M. 2013. *Lifeblood: Oil, Freedom, and the Forces of Capital*. Minneapolis: University of Minnesota Press.

Hughes, D.M. 2017. *Energy without Conscience: Oil, Climate Change, and Complicity*. Durham: Duke University Press.

IEEP. 2020. *More than Half of All CO2 Emissions since 1751 Emitted in the Last 30 Years*. Brussels: Institute for European Environmental Policy.

Illich, I. 2013. The Social Construction of Energy. In *Beyond Economics and Ecology: The Radical Thought of Ivan Illich*, ed. S. Samuel, 105–123. London: Marion Boyars.

Ishiyama, N., and K. TallBear. 2001. Changing Notions of Environmental Justice in the Decision to Host a Nuclear Storage Facility on the Skull Valley Goshute Reservation. In *Session 51: Equity and Environmental Justice*. Presented at the Waste Management 2001 Symposia, Tucson, AZ.

Jalbert, K., A. Willow, D. Casagrande, and S. Paladino, eds. 2017. *ExtrACTION: Impacts, Engagements and Alternative Futures*. New York: Routledge.

Jarratt-Snider, K., and M.O. Nielsen, eds. 2020. *Indigenous Environmental Justice*. Tucson: University of Arizona Press.

Jenkins, K. 2018. Setting Energy Justice Apart from the Crowd: Lessons from Environmental and Climate Justice. *Energy Research & Social Science* 39: 117–121.

Johnson, B. 2016. Energy Slaves: Carbon Technologies, Climate Change, and the Stratified History of the Fossil Economy. *American Quarterly* 68 (4): 955–979.

———. 2019. *Mineral Rites: An Archaeology of the Fossil Economy*. Baltimore: Johns Hopkins University Press.

Kallis, G., S. Paulson, G. D'Alisa, and F. Demaria. 2020. *The Case for Degrowth*. Cambridge: Polity Press.

Kenrick, J. 2013. Emerging from the Shadow of Climate Change Denial. *ACME: An International E-Journal for Critical Geographies* 12 (1): 102–130.

Kojola, E., and D.N. Pellow. 2021. New Directions in Environmental Justice Studies: Examining the State and Violence. *Environmental Politics* 30 (1–2): 100–118.

Kothari, A. 2014. Radical Ecological Democracy: A Path Forward for India and Beyond. *Development* 57 (1): 36–45.

———. 2019. Radical Well-being Alternatives to Development. In *Research Handbook on Law, Environment and the Global South*, ed. P. Cullet and S. Koonan, 64–85. Cheltenham: Edward Elgar.

Kurtz, H.E. 2010. Acknowledging the Racial State: An Agenda for Environmental Justice Research. In *Spaces of Environmental Justice*, ed. R. Holifield, M. Porter, and G. Walker, 95–115. Chichester: Wiley-Blackwell.

LaDuke, W. 1999. *All Our Relations: Native Struggles for Land and Life*. Cambridge/Minneapolis: South End Press/Honor the Earth.

Lander, E. 2013. Complementary and Conflicting Transformation Projects in Heterogeneous Societies. In *Beyond Development: Alternative Visions from Latin America*, ed. M. Lang and D. Mokrani, 87–104. Quito: Transnational Institute.

LeMenager, S. 2014. *Living Oil: Petroleum Culture in the American Century*. Oxford University Press.

Lennon, M. 2017. Decolonizing Energy: Black Lives Matter and Technoscientific Expertise Amid Solar Transitions. *Energy Research & Social Science* 30: 18–27.

———. 2021. Energy Transitions in a Time of Intersecting Precarities: From Reductive Environmentalism to Antiracist Praxis. *Energy Research & Social Science* 73: 101930.

Lohmann, L., and N. Hildyard. 2014. *Energy, Work and Finance*. Dorset: The Corner House.

Loloum, T., S. Abram, and N. Ortar, eds. 2021. *Ethnographies of Power: A Political Anthropology of Energy*. New York: Berghahn Books.

Luke, N., and N. Heynen. 2020. Community Solar as Energy Reparations: Abolishing Petro-Racial Capitalism in New Orleans. *American Quarterly* 72 (3): 603–625.

Mahmud, T. 2013. Cheaper Than a Slave: Indentured Labor, Colonialism and Capitalism. *Whittier Law Review* 34 (2): 215.

Maldonado, J.K. 2019. *Seeking Justice in an Energy Sacrifice Zone: Standing on Vanishing Land in Coastal Louisiana*. New York: Routledge.

Malm, A. 2016. *Fossil Capital: The Rise of Steam Power and the Roots of Global Warming*. London: Verso.

Maris, C. 1991. Milking the Meter: On Analogy, Universalisability and World Views. In *Legal Knowledge and Analogy: Fragments of Legal Epistemology, Hermeneutics and Linguistics*, ed. P. Nerhot, 71–106. Dordrecht: Springer.

Marks, R. 2002. *The Origins of the Modern World: A Global and Environmental Narrative from the Fifteenth to the Twenty-First Century*. Lanham: Rowman & Littlefield.

Martinez-Alier, J. 2002. *The Environmentalism of the Poor: A Study of Ecological Conflicts and Valuation*. Cheltenham: Edward Elgar.

Martinez-Alier, J., I. Anguelovski, P. Bond, D.D. Bene, F. Demaria, J.-F. Gerber, L. Greyl, W. Haas, H. Healy, V. Marín-Burgos, G. Ojo, M. Porto, L. Rijnhout, B. Rodríguez-Labajos, J. Spangenberg, L. Temper, R. Warlenius, and I. Yánez. 2014. Between Activism and Science: Grassroots Concepts for Sustainability Coined by Environmental Justice Organizations. *Journal of Political Ecology* 21: 19–60.

Martinez-Alier, J., L. Temper, D. Del Bene, and A. Scheidel. 2016. Is There a Global Environmental Justice Movement? *The Journal of Peasant Studies* 43 (3): 731–755.

Marx, K., and F. Engels. 1975. *Collected Works, vol. 46*. New York: International Publishers.

Mbembe, A. 2004. Aesthetics of Superfluity. *Public Culture* 16 (3): 373–405.

McGregor, D., S. Whitaker, and M. Sritharan. 2020. Indigenous Environmental Justice and Sustainability. *Current Opinion in Environmental Sustainability* 43: 35–40.

McKittrick, K. 2006. *Demonic Grounds: Black Women and the Cartographies of Struggle*. Minneapolis: University of Minnesota Press.

———. 2013. Plantation Futures. *Small Axe: A Caribbean Journal of Criticism* 17 (3): 1–15.

McNeill, J.R. 2000. *Something New Under the Sun: An Environmental History of the Twentieth-Century World*. New York: W. W. Norton.

Melamed, J. 2015. Racial Capitalism. *Critical Ethnic Studies* 1 (1): 76–85.

Mellor, M. 2006. Ecofeminist Political Economy. *International Journal of Green Economics* 1 (1/2): 139–150.

Mendoza, M., M. Greenleaf, and E.H. Thomas. 2021. Green Distributive Politics: Legitimizing Green Capitalism and Environmental Protection in Latin America. *Geoforum* 126: 1–12.

Merchant, C. 1990. *The Death of Nature: Women, Ecology, and the Scientific Revolution*. New York: HarperOne.

Mies, M., and V. Shiva. 1993. *Ecofeminism*. London: Zed Books.

Mitcham, C., and J.S. Rolston. 2013. Energy Constraints. *Science and Engineering Ethics* 19 (2): 313–319.

Mitchell, T. 2011. *Carbon Democracy: Political Power in the Age of Oil*. London: Verso.

Mookerjea, S. 2017. Petrocultures in Passive Revolution: The Autonomous Domain of Treaty Poetics. In *Petrocultures: Oil, Politics, Culture*, ed. S. Wilson, A. Carlson, and I. Szeman, 325–354. Montreal; Chicago: McGill-Queen's University Press.

Mouhot, J.-F. 2011. Past Connections and Present Similarities in Slave Ownership and Fossil Fuel Usage. *Climatic Change* 105 (1–2): 329–355.

Müller, B. 2019. The Careless Society: Dependency and Care Work in Capitalist Societies. *Frontiers in Sociology* 3 (44): 1–10.

Mulvaney, D. 2019. *Solar Power: Innovation, Sustainability, and Environmental Justice*. Oakland: University of California Press.

Munro, P., G. van der Horst, and S. Healy. 2017. Energy Justice for All? Rethinking Sustainable Development Goal 7 through Struggles over Traditional Energy Practices in Sierra Leone. *Energy Policy* 105: 635–641.

Newell, P. 2021. Race and the Politics of Energy Transitions. *Energy Research & Social Science* 71: 101839.

Newell, P., and D. Mulvaney. 2013. The Political Economy of the 'Just Transition'. *The Geographical Journal* 179 (2): 132–140.

Newell, P., and M. Paterson. 2010. *Climate Capitalism: Global Warming and the Transformation of the Global Economy*. Cambridge University Press.

Nikiforuk, A. 2012. *The Energy of Slaves: Oil and the New Servitude.* Vancouver: Greystone Books.

Nonini, D. 2006. The Global Idea of 'the Commons'. *Social Analysis: The International Journal of Social and Cultural Practice* 50 (3): 164–177.

Nye, D. 1998. *Consuming Power: A Social History of American Energies.* The MIT Press.

Omi, M., and H. Winant. 1994. *Racial Formation in the United States: From the 1960s to the 1990s.* New York: Routledge.

de Onís, C.M. 2021. *Energy Islands: Metaphors of Power, Extractivism, and Justice in Puerto Rico.* Oakland: University of California Press.

Park, L.S.-H., and D. Pellow. 2011. *The Slums of Aspen: Immigrants vs. the Environment in America's Eden.* NYU Press.

Partridge, T. 2017a. Unconventional Action and Community Control: Rerouting Dependencies Despite the Hydrocarbon Economy. In *ExtrACTION: Impacts, Engagements and Alternative Futures,* ed. K. Jalbert, A. Willow, D. Casagrande, and S. Paladino, 198–210. New York: Routledge.

———. 2017b. Resisting ruination: resource sovereignties and socioecological struggles in Cotopaxi, Ecuador. *Journal of Political Ecology* 24, 763–776.

———. 2018. The commons as organizing infrastructure: Indigenous collaborations and post-neoliberal visions in Ecuador. In *The Right to Nature: Social movements, environmental justice and neoliberal natures,* eds. E. Apostolopoulou and J. Cortes-Vazquez, 251–262. London: Routledge.

———. forthcoming. The Right to Energy: Learning from Struggles for Food, Water, and Rights to Nature. In *Handbook on Energy Justice,* ed. S. Bouzarovski, S. Fuller, and T. Reames. Cheltenham: Edward Elgar.

Partridge, T., M. Thomas, N. Pidgeon, and B.H. Harthorn. 2018. Urgency in Energy Justice: Contestation and Time in Prospective Shale Extraction in the United States and United Kingdom. *Energy Research & Social Science* 42: 138–146.

Pellow, D.N. 2001. Environmental Justice and the Political Process: Movements, Corporations, and the State. *The Sociological Quarterly* 42 (1): 47–67.

———. 2018. *What is Critical Environmental Justice?* Cambridge: Polity Press.

POCELS. 1991. *The Principles of Environmental Justice, drafted and adopted by Delegates to the First National People of Color Environmental Leadership Summit.* October 24–27: Washington, DC.

Polletta, F. 2002. *Freedom is an Endless Meeting: Democracy in American Social Movements.* University of Chicago Press.

Powell, D.E. 2015. The Rainbow is Our Sovereignty: Rethinking the Politics of Energy on the Navajo Nation. *Journal of Political Ecology* 22: 53–78

———. 2018. *Landscapes of Power: Politics of Energy in the Navajo Nation.* Durham: Duke University Press.

Pritchard, S. 2015. Situating 'Routes of Power' within the History of Technology. *H-Environment Roundtable Reviews* 5 (9): 14–23.

Pulido, L. 2017. Geographies of Race and Ethnicity II: Environmental Racism, Racial Capitalism and State-Sanctioned Violence. *Progress in Human Geography* 41 (4): 524–533.

Quaempts, E.J., K.L. Jones, S.J. O'Daniel, T.J. Beechie, and G.C. Poole. 2018. Aligning Environmental Management with Ecosystem Resilience: A First Foods Example from the Confederated Tribes of the Umatilla Indian Reservation, Oregon, USA. *Ecology and Society* 23 (2): art29.

Ranco, D.J. 2008. The Trust Responsibility and Limited Sovereignty: What can Environmental Justice Groups Learn from Indian Nations? *Society & Natural Resources* 21 (4): 354–362.

Ranco, D.J., C.A. O'Neill, J. Donatuto, and B.L. Harper. 2011. Environmental Justice, American Indians and the Cultural Dilemma: Developing Environmental Management for Tribal Health and Well-being. *Environmental Justice* 4 (4): 221–230.

Razack, S.H. 2002. *Race, Space, and the Law: Unmapping a White Settler Society.* Toronto: Between the Lines.

Robinson, C. 1983. *Black Marxism: The Making of the Black Radical Tradition.* Chapel Hill: University of North Carolina Press.

Rocheleau, D., B. Thomas-Slayter, and E. Wangari, eds. 1996. *Feminist Political Ecology: Global Issues and Local Experiences.* New York: Routledge.

Rodgers, D., and B. O'Neill. 2012. Infrastructural Violence: Introduction to the Special Issue. *Ethnography* 13 (4): 401–412.

Roy, B. 2022. The Connecting Thread: How Climate Activists in South Asia Built Global Solidarity against Coal. In *Property Will Cost Us the Earth: Direct Action and the Future of the Global Climate Movement*, ed. J. Kindig, 66–71. London: Verso.

Santiago, R., H. Lloréns, C. Garcia-Quijano, and C. de Onís 2017. Dispatch from the Frontlines of Puerto Rico in a Post-Maria World. *Latino Rebels*, 9 October.

Santos, B. de S. 2007. *The Rise of the Global Left: The World Social Forum and Beyond.* London: Zed Books.

Scott, J.C. 2010. *The Art of Not Being Governed: An Anarchist History of Upland Southeast Asia.* New Haven: Yale University Press.

Simpson, A. 2014. *Mohawk Interruptus: Political Life across the Borders of Settler States.* Durham: Duke University Press.

Simpson, L.B. 2021. *A Short History of the Blockade: Giant Beavers, Diplomacy, and Regeneration in Nishnaabewin.* Edmonton: University of Alberta Press.

Smil, V. 2004. World History and Energy. In *Encyclopedia of Energy*, 549–561. Elsevier.

Speed, S. 2017. Structures of Settler Capitalism in Abya Yala. *American Quarterly* 69 (4): 783–790.

Stephens, J.C. 2019. Energy Democracy: Redistributing Power to the People Through Renewable Transformation. *Environment: Science and Policy for Sustainable Development* 61 (2): 4–13.

Svampa, M. 2019. *Las fronteras del neoextractivismo en América Latina: conflictos socioambientales, giro ecoterritorial y nuevas dependencias.* Bielefeld University Press.

Sydee, J., and S. Beder. 2001. Ecofeminism and Globalisation: A Critical Appraisal. *Democracy & Nature* 7 (2): 281–302.

Sze, J. 2005. Race and Power: An Introduction to Environmental Justice Energy Activism. In *Power, Justice, and the Environment: A Critical Appraisal of the Environmental Justice Movement,* ed. D.N. Pellow and R.J. Brulle, 101–115. The MIT Press.

Szeman, I., and D. Boyer, eds. 2017. *Energy Humanities: An Anthology.* Baltimore: Johns Hopkins University Press.

TallBear, K. 2019. Badass Indigenous Women Caretake Relations: #Standingrock, #IdleNoMore, #BlackLivesMatter. In *Standing with Standing Rock: Voices from the #NoDAPL Movement,* ed. N. Estes and J. Dhillon, 13–18. University of Minnesota Press.

Trouillot, M.-R. 2003. *Global Transformations: Anthropology and the Modern World.* New York: Palgrave.

Tsosie, R.A. 2007. Indigenous People and Environmental Justice: The Impact of Climate Change. *University of Colorado Law Review* 78: 1625–1677.

UN. 2016. *Goal 7: Ensure Access to Affordable, Reliable, Sustainable and Modern Energy for All—SDG Indicators.* New York: United Nations Statistics Division.

Veltmeyer, H., and J. Petras. 2014. *The New Extractivism: A Post-Neoliberal Development Model or Imperialism of the Twenty-First Century?* London: Zed Books.

Weaver, J., ed. 1996. *Defending Mother Earth: Native American Perspectives on Environmental Justice.* New York: Orbis.

Weeks, K. 2011. *The Problem with Work: Feminism, Marxism, Antiwork Politics, and Postwork Imaginaries.* Durham: Duke University Press.

Weheliye, A.G. 2014. *Habeas Viscus: Racializing Assemblages, Biopolitics, and Black Feminist Theories of the Human.* Durham: Duke University Press.

White, L. 1959. *The Evolution of Culture: The Development of Civilization to the Fall of Rome.* New York: McGraw-Hill.

White, R. 1996. *The Organic Machine: The Remaking of the Columbia River.* New York: Hill & Wang.

Whyte, K.P. 2017a. The Dakota Access Pipeline, Environmental Injustice, and U.S. Colonialism. *Red Ink: An International Journal of Indigenous Literature, Arts, & Humanities* 19 (1): 154–169.

———. 2017b. The Recognition Paradigm of Environmental Injustice. In *The Routledge Handbook of Environmental Justice,* ed. R. Holifield, J. Chakraborty, and G. Walker, 113–123. London: Routledge.

———. 2018. Settler Colonialism, Ecology, and Environmental Injustice. *Environment and Society* 9 (1): 125–144.

———. 2020. Too Late for Indigenous Climate Justice: Ecological and Relational Tipping Points. *Wiley Interdisciplinary Reviews: Climate Change* 11 (1). https://doi.org/10.1002/wcc.603.

Williams, E.E. 2014. *Capitalism and Slavery.* Chapel Hill: University of North Carolina Press.

Willow, A. 2018. *Understanding ExtrACTIVISM: Culture and Power in Natural Resource Disputes.* London: Routledge.

Wrigley, E. 1988. *Continuity, Chance and Change: The Character of the Industrial Revolution in England.* Cambridge University Press.

Yazzie, M.K. 2018. Decolonizing Development in Diné Bikeyah: Resource Extraction, Anti-Capitalism, and Relational Futures. *Environment and Society* 9 (1): 25–39.

Young, D. 2019. Rethinking the Shotgun Marriage of Freud and Marx: Monetary Subjects without Money, the Socialization of the Death Drive, and the Terminal Crisis of Capitalism. *Mediations* 32 (2): 99–138.

Yuen, E. 2001. Introduction. In *The Battle of Seattle: The New Challenge to Capitalist Globalization,* ed. E. Yuen, D. Burton Rose, and G. Katsiaficas. New York: Soft Skull Press.

Yusoff, K. 2018. *A Billion Black Anthropocenes or None.* Minneapolis: University of Minnesota Press.

Ziser, M., N. Zaretsky, and J. Sze. 2020. Perpetual Motion: Energy and American Studies. *American Quarterly* 72 (3): 543–557.

Zografos, C. 2019. Direct Democracy. In *Pluriverse: A Post-development Dictionary,* ed. A. Kothari, A. Salleh, A. Escobar, F. Demaria, and A. Acosta, 154–157. New Delhi: Tulika Books.

CHAPTER 2

Transitions Beyond Crisis: Pluralism, Restoration, Degrowth

Abstract EJ movements and analyses play vital roles in building future energy systems that are more equitable, less destructive, and which support a just transition away from global dependency on fossil fuels. Contemporary energy debates are predominantly framed in terms of crises and transitions. Yet the meaning and purposes of transitions remain contested, as do questions concerning who controls, directs, and implements transitions. Movements already fighting for transformative social change show how different transitions projects redesign relations between energy systems, ecosystems, communities, and society as a whole. Insights from these multiple, global movements further underline the importance of creating transitions not just to different energy systems but to different ways of living and relating. Studying transitions therefore involves analyzing the socio-environmental relations that different energy choices support and reinforce or, by contrast, devalue and undermine. This section highlights the importance for transitions practice and analysis of four key concepts (pluralism, diversity, restoration, and degrowth), viewing transitions as more than a response to crisis. The result is a reframing of energy transitions as collaborative projects of justice; that is, as socio-political processes where collective action, centered on consent and reciprocity, is allowed to thrive.

Keywords Energy transitions • Pluralism • Diversity • Restorative environmental justice • Degrowth • Energy sovereignty

© The Author(s), under exclusive license to Springer Nature
Switzerland AG 2022
T. Partridge, *Energy and Environmental Justice*,
https://doi.org/10.1007/978-3-031-09760-7_2

INTRODUCTION: TRANSITIONS AS COLLABORATIVE PROJECTS OF JUSTICE

Contemporary energy debates are predominantly framed in terms of crises and transitions. Devastating environmental damage and climate change—a result of massive deforestation and the increasing intensity of atmospheric CO_2 caused primarily by energy use—has seen the idea of a transition away from fossil fuel dependency to renewable energy sources become a standard feature of political discourse (Stirling 2014). The meaning and purposes of transitions, however, are not always clear.

The phrase 'energy transition' implies both the development of renewable energy infrastructures *and* a decline in the use of established energy sources, yet recent growth in renewable energy is effectively adding to global energy production rather than replacing fossil fuels (Mulvaney 2019; York and Bell 2019). Over the years following the first of the UN climate summits—COP1 in Berlin in 1995—total annual CO_2 emissions worldwide have increased approximately 60% (Malm 2021, p. 11). Cases of 'greenwashing' abound, where energy policies that do little or nothing to address demand or to phase-out fossil fuel extraction are reclassified as an instrumental part of the 'energy transition.' In March 2022, the UK Government "Oil and Gas Authority" rebranded as the "North Sea Transition Authority," a move that the chief executive accompanied with the statement, "our values remain the same" [1]—just as the Government is expected to approve extraction at six new oil and gas fields in the North Sea. [2]

Ongoing growth in global energy demand—energy which is consumed very unevenly across the world's regions—undermines any implied, or substitutive, transition away from hydrocarbons to low-carbon energy sources. Increasing rates of energy demand are now projected to outpace increasing low-carbon energy capacity. The world produces 8 billion more megawatt hours of clean energy than in the year 2000, but energy demand has grown by 48 billion megawatt hours over the same period, thus new clean energy capacity covers "only 16% of new demand" (Hickel 2019,

---

[1] Source: "Press release: Oil and Gas Authority changes name to North Sea Transition Authority, 21 March 2022", as at: https://www.nstauthority.co.uk/news-publications/news/2022/oil-and-gas-authority-changes-name-to-north-sea-transition-authority/ Accessed: 30 March 2022.

[2] Source: https://www.independent.co.uk/climate-change/news/oil-gas-fields-net-zero-b2010060.html Accessed: 29 March 2022.

p. 55). If the 'energy transition' is to actually reduce social and environmental harm on a significant scale, then the transition has to instigate change beyond the substitution of some sources of energy for others.

As described below, many of the social movements whose work contributed to the global spread of the energy transition concept began with a focus on deep social change rather than (only) the reconfiguration of energy systems. Much of that original movement focus has since been overshadowed as the scale of plans for the energy transition have grown and those plans taken over by state and corporate actors. While the scale, and depth, of changes necessary are undeniable, it matters greatly how those changes are imagined and implemented (and by who). While different versions of the US Green New Deal (GND) represent bold proposals to transform and decarbonize Global North economies in order to address climate change, EJ scholars and activists highlight the risk of unevenly burdening low-carbon economies with the social, environmental, health, and economic costs involved (Zografos and Robbins 2020). Activists continue to push for GND initiatives to include climate justice, reductions in excess and overall consumption, and protections for ecological and worker wellbeing across the Global South (Ordoñez Muñoz 2021). Alongside calls for an Indigenous-led GND (LaDuke 2022), the Red Nation collective push for The Red Deal which incorporates fulfilling Land Back claims, foregrounds decolonization and an end to border imperialism, and replacing the dominant 'militarized extractive economy' with a 'caretaking economy' centered on land-based movements and all forms of care work in protection of people, land, water, and treaty rights (The Red Nation 2021, p. 28f). EJ and Indigenous movements continue to lead the way in defining what 'transitions' need to involve.

A standard policy focus on important metrics—renewable energy output and atmospheric carbon—obscures what historians of energy have repeatedly shown: energy transitions past and present are about much more than energy sources and are enmeshed in processes of social, economic, political, and environmental change (e.g., Strauss et al. 2013; Jones 2014; Malm 2016). It is therefore a mistake to focus only on energy systems and practices: transitions also present opportunities to rethink dominant political and economic institutions (Howe and Boyer 2016). EJ scholarship and activism offer vital tools for radically re-assessing the notion of 'transition'—rejecting responses that overlook the need for structural change while actively recognizing (and facilitating) the diverse ways in which transitions can be transformative.

What if we were to reframe transitions as collaborative projects of justice? That is, to see transitions as socio-political processes where collective action is allowed to thrive and build toward environmental protection, equity, and the abolition of racial capitalism. What theoretical, practical, or political spaces are opened up, what radical possibilities are revealed or reaffirmed? What might the consequences be for demanding and enacting multiple forms of justice? Each collaborative project of justice—which, as the preceding sections have underlined, would be centered on consent and reciprocity—would be more than a response to crisis. Instead, such projects seek to redesign relations between energy systems, communities, and society as a whole. Reframing and re-centering approaches to transitions along these lines is done by analyzing the socio-environmental relations that different energy choices strengthen or, by contrast, devalue and undermine. Drawing these threads and connections together, the remaining sections of Chap. 2 highlight the importance for energy transitions of four key concepts: pluralism, diversity, restoration, and degrowth.

In India, the growing movement for Radical Ecological Democracy (or *eco-swaraj*) draws on principles in the writings of Gandhi and others in order to explore new ways of combining political, economic, cultural, and ecological relationships—particularly by relocating power within collectives and communities rather than with states and corporations (Gandhi 1909; Kothari 2014; Shrivastava 2017; Partridge 2020). Eco-swaraj thus seeks to address collective and individual wellbeing in part by ensuring everyone has the right, capacity, and opportunity to take part in decision-making processes (Kothari et al. 2019). Through these principles, eco-swaraj aligns with a diverse, global range of transformative initiatives that support a shift from the sense of a universe to the pluriverse—creating a world that contains a multiplicity of possible worlds and alternatives (ibid.). The pluriverse concept presents critical questions for how we might understand and implement different kinds of transition:

"Are the means of economic production and social reproduction justly controlled? Are humans relating to non-humans in mutually enhancing ways? Do all people have access to meaningful livelihoods? Is there a just intergenerational distribution of bads and goods? Are traditional or modern discriminations of gender, class, ethnicity, race, caste, and sexuality being erased? Are peace and non-violence infused throughout community life?". (Kothari et al. 2019, p. xix)

These questions push beyond "reformist-incrementalist" approaches to transitions in order to see how "disruptive and systemic-structural socio-energy transformations" can help create the radical social change that this moment requires (Scoones et al. 2015, cited by Healy and Barry 2017, p. 452). Truly sustainable transitions would also have to attend to our shared capacity for joy—a future worth saving is a future enjoyed by everybody (Vemuri and Barney 2022). These would be transitions not just to different energy systems but to different ways of living and relating.

Pluralism and Diversity

EJ work recognizes that the impacts of unsustainable resource consumption around the world are not experienced equally: poor and powerless people are denied access to healthy environments (and protection from environmental harms) in ways that those with economic resources and political power are not, and those inequalities are profoundly shaped by ongoing histories of racism and domination (Pellow and Brulle 2005, pp. 2–3). For energy transitions to play a role in addressing these central EJ concerns, simultaneous work of dismantling and creativity is required—taking action to dissolve the social structures and institutions of power that replicate widespread relations of inequality while also enacting social and political changes that support the realization of environmental justice within people's lived experiences (Temper et al. 2018). Rooting the idea of transition within diverse individual experiences expands the terms and goals of transitions: both confronting the unequal effects of structures of control and systems of production while also considering how environmental harms differentially affect individual bodies. This includes recognizing diverse embodiments and different experiences of disablement and disability (Jaffee and John 2018). As Gutierrez et al. (2021, p. 78) emphasize, recent work in queer theory, disability studies, and environmental humanities is focusing EJ on how toxicity creates different bodily experiences not only across but also within contexts (Johnson 2017; Ray and Sibara 2017). Pluralism and diversity are the basis of transitions as collaborative projects of justice, offering starting points and strategies (not merely desired goals) for collective action.

Transitions based on engagement with EJ research and activism are plural. One collective approach that started in Europe—the Transition Towns Movement, which grew from a permaculture project to (re)imagine energy futures, published in 2005 as "Kinsale 2021: an Energy Descent

Action Plan"—encourages creativity in exploring place-specific adaptations of suggested processes and principles for deepening community resilience (Hopkins 2008). Suggested principles include: localizing production of food, energy, and building materials wherever possible; designing projects and enterprises that are low carbon with regard to both inputs and outputs; bringing assets (such as land, businesses, energy generation, buildings) into community ownership; creating a vision of abundance for the future while recognizing that our world is one in which credit, resources, and the materials that support energy systems are finite; and endorsing business models that are not purely for personal profit (such as social enterprises and cooperatives) (Hopkins 2008, 2019). Tailored to the specific contours of social life and ecosystems in different contexts, the variety of ways of implementing these ideas is practically limitless. That said, the notion of 'energy descent' remains central to these diverse forms of transition.

Building diverse, place-specific transitions recognizes energy as a "material-discursive force" which, whenever it is put to use, reflects ideas about what matters and how diverse relationships are to be organized, controlled, and valued (Lennon 2017, p. 24). Put differently, this is to emphasize the interdependence of physical and social infrastructures as well as the outsize influence that energy practices have in mediating those connections (Partridge 2017, 2018). By engaging with human and non-human needs in different social and geographical contexts, transitions can be constructed in such a way that they better enmesh with the biodiversity and complex ecosystem relations that already support life on Earth. Transition projects and movements are therefore inherently pluralistic in their orientation, which is itself a mode of engagement with socioecological complexity and diversity. In the words of Gustavo Esteva (2005, p. 153): "the only legitimate, coherent and sensible attitude before the real plurality of the world is radical pluralism." The emphasis here is on *radical*—in order to avoid replicating state-based assimilationist models of 'liberal pluralism' which, in Canada and other settler states, force Indigenous identities to fit within state-sanctioned groups, institutions, and narratives (Walia 2012). A core challenge for energy transitions, then, is to reconfigure energy practices such that they contribute to building less destructive, more equitable social relationships within complete and complex ecosystems and within challenging political contexts.

EJ and Transition Town Movement perspectives emphasize this importance of tailoring energy systems to particular political contexts (in ways that support more democratic systems of energy governance) while also, crucially, recognizing the diversity of lived experiences *within* particular social and political settings. At the same time, transitions thinking and action has to address countervailing forces, specifically by identifying and resisting structural conditions that exacerbate experiences of injustice (and which limit or reduce support for diverse, democratic initiatives). There are many groups who are routinely marginalized within human societies, including "women, immigrants, LGBTQ persons, people of color, indigenous peoples, disabled persons, the elderly, children, low-income people, and nonhuman species.

> And while the *experiences* of these various groups are qualitatively distinct (they are not equivalent), the *logic* of domination and othering as practiced by more powerful groups against them provides a thread of intersectionality through each of their oppressions". (Pellow 2018, pp. 28–29, original emphasis)

A core issue here is not only that people may experience multiple forms of oppression but also that oppressive forces interact and make those combined, intersectional experiences of injustice more acute. Intersectional analyses are in affinity with the work of people acting and writing within women-of-color feminism, work that shows how "procedures of race, class, gender, and sexuality and economic forces aggregate and interlock to create the lived conditions of the everyday" (Melamed 2011, p. 226). Applying these approaches from critical environmental justice to transitions thinking, two connected issues come to the fore: (i) the need to recognize the persistence of forces of political and economic domination (identifying the structural relations that need to be dismantled and reimagined) and (ii) the need to recognize how diversity and pluralism are fundamental to deepening community resilience and social justice.

Non-local scholars and activists do a great disservice to affected EJ communities when they apply universal perspectives and neglect the unevenness of different people's experiences of environmental injustice, not only from one community to the next but also (and especially) when that unevenness is experienced within particular sociocultural contexts (Powell 2018). Gathering together accounts of uranium mining from Navajo people, Brugge et al. emphasize the absence of a single internally consistent position among witnesses and underline how community diversity creates a wide variety of conceptions of what would constitute justice

(Brugge et al. 2006). Within ongoing debates about "transition" in Diné territory, different kinds of energy future and definitions of justice are reflected in the goals of different groups: "'transition' to the Just Transition Coalition of environmental activists means something rather different from "transition" to the Navajo Transitional Energy Corporation... while the Just Transition Coalition seeks a movement away from a fossil fuel-based economy through proposals that were decolonizing (requiring California utilities to subsidize Navajo solar and wind projects), NTEC's 'transition' signals a new modality of power in ownership while maintaining dependence on coal" (Powell 2018, p. 94). Respecting multiple definitions of justice, even or especially when they include difficult or apparently contradictory goals, prepares the ground for actions that are more responsive to needs that may change over time and actions which disrupt externally-imposed, purportedly universal, proposals.

Countering inaccurate generalizations in common statements about Indigenous peoples and Indigenous knowledge in relation to climate change—statements that tend to focus only on demise and vulnerability rather than also highlighting different adaptive capabilities and forms of resilience—Maria Bargh argues for greater recognition of the multiple roles and identities that Indigenous peoples inhabit, then applies this to transition thinking more broadly: "Many people simultaneously inhabit positions of employer, employee, unemployed, mother, friend, neighbor, land owner, consumer—which give rise to different economic interactions. People are never only *just* ecologists or workers or capitalists" (Bargh 2021, p. 422, original emphasis). Drawing on J. K. Gibson-Graham's diverse economies framework, the emphasis here is on recognizing, and developing, the collective potential that is held in the multiple roles people hold individually and collectively (Gibson-Graham and CEC 2017; Bargh 2021). By articulating multiple economic roles within new energy systems—beyond being merely consumers or producers whose roles are over-determined by calculations of supply and demand—opportunities are created to rethink possible scales of collective action and to prioritize solidarity and connection (Coleman 2021). Such context-specific, place-based approaches to transitions are non-prescriptive: there is no direct model that can be applied universally.

Methodologically, non-prescriptive approaches to transitions thinking require flexibility and responsiveness in their design and implementation. These are actions that reconfigure diverse resources—whether material or primarily social, based in goods or in relationships, perspectives, purposes, and values—in ways that help communities avoid some of the social, economic and environmental costs of extractive energy systems and to build collective strategies to withstand the forms of uninvited change those systems create (Partridge 2017). Future-oriented, localized actions of this kind reassess how different energy practices are avoided or integrated into everyday life. These approaches orientate 'transition' initiatives around the need to 'dig where you stand' (Maguire 1987). Against a dominant position that places hope in the application of novel technological interventions as a way to transition to modified energy systems, digging where you stand means working within extant communities and within the complex webs of socioecological relationships that have been successively degraded by extractive and exploitative social systems. Pursuing such actions often requires a degree of political autonomy in how decisions are made in order to address, in more equitable and participatory ways, who is invited to take on roles of leadership within decision-making processes.

This aligns closely with the Transition Towns concept of an 'Inner Transition' which invites groups to find further ways of avoiding the replication of methods, models, or strategies that ultimately cause harm and which undermine the basis of effective, careful cooperation. The 'Inner Transition' recognizes that "'how' a group does a project matters as much as 'what' they do… Working in a way that prioritizes paying attention to the factors that lead to burnout [and] develops vital skills around making decisions, running effective meetings, managing conflicts, and so on" (Hopkins 2019, p. 318). By attending to different needs within the group, such prefigurative action reaffirms the importance of pluralism within transitions thinking.

Analytically, pluralist approaches to transitions echo pluralist approaches to the study and pursuit of justice. These complement an established sociological focus on empirical aspects of social struggles (such as structures, frames, resources, mobilization, organization, and contention) by also addressing normative questions, especially those that concern the contested *meanings* of different rights and justice claims (Somers 2008, p. xiii). Such a pluralist 'knowledge culture' recognizes that these two dimensions—the empirical and the normative—are inextricably linked and are 'mutually interdependent' (ibid.). This means that "what is normatively held to be just is not determined in isolation from historical and existing social arrangements, social hierarchies, and differentials in social power… what *is* just is not determined by beginning with abstract principles of justice derived in isolation from extant social formations, as is characteristic of liberal, Western moral philosophy following from Rawls. Rather, what *is* normatively just becomes determined through the course of social interaction, communication, and conflict in empirical, historical, and political-economic contexts" (van Gerven 2022, p. 26, original emphasis). For transitions to become collaborative projects of justice, then, a parallel approach is required: group goals (the kinds of social, economic, and energy relations that *should* be implemented) are defined and redefined throughout the collective (empirical) processes of working toward them. The objectives of prefiguration evolve over time through the negotiation of shared needs and actions.

Temporally, pluralism in transitions thinking also extends to the discourse and framings of 'crisis,' which are not universally understood or applicable. Reviewing the 2012–2017 Enbridge Line 9 oil pipeline dispute in Canada's Great Lakes region, Awâsis (2020) describes how Indigenous ways of living that embrace multiple temporalities—and which acknowledge overlapping understandings of epochs, timeframes, and kinship relations—resist environmental reviews that focus only on short-term impacts of energy infrastructures (reviews which thus also disregard non-human temporalities). In specific cases such as this, as well as in broader considerations of energy transition, imposed timelines and timeframes represent another form of harm. A lot of transitions discourse (including within EJ) is synchronic, applying a diagnostic framework of justice principles at a particular moment in time in order to propose improvements to contemporary energy practices. Synchronic approaches operate through the language of crisis, problems, and solutions—recognizing the urgency of injustice but failing to support the radical, structural change that is required to avoid replicating colonial power and violence against marginalized groups (Whyte 2021). This is a form of analytical 'presentism' and it has the effect of not only amplifying a discourse of newness and crisis but also failing to recognize how such actions obscure past experiences of violence (Whyte 2021). Dominant, methodological presentism further entrenches an institutional perspective on social change, minimizing the influence of diverse resistance movements and assuming, falsely, that extant political systems can deliver justice for Indigenous and marginalized groups (when those political systems routinely denigrate and discriminate against those same people and communities).

Referencing diverse Native experiences, synchronic approaches—and the accompanying discourse of newness and crisis—are further critiqued with regard to 'apocalyptic' thinking. This involves a rejection of the dominant imaginary of a coming apocalypse—something that requires 'urgent' action to forestall—because Indigenous peoples have inhabited a 'postapocalyptic world' since the invasions of colonialism began, even while renewed relationships emerge through resistance and kin-making (Dillon 2012; Gross 2014; Powell 2018; Whyte 2020). In terms of pluralism, what these views make clear is that dominant 'crisis' discourse is problematic both because it imposes a singular temporal frame for understanding current conditions of degradation and injustice *and* because it assumes a universal experience of those conditions.

Even in contexts where the frame of multiple crises is widely accepted, what actually constitutes different types of crisis (and how they are identified and differentially experienced) varies significantly across different global contexts. The model of social life in crisis in Europe, for example, is marked by state withdrawal and the crises of advanced capitalism, while in Latin America the model is dominated by extractivist policies and the precarity that comes from dependence on raw material exports to the global market (Escobar 2015). Transitions-beyond-crisis are society-building projects: it matters greatly who leads, works on, and contributes to such processes (Partridge forthcoming-b). Together, these many distinct experiences and critiques carry a key lesson for energy transitions thinking: the need for sensitivity to specific unfolding histories that disrupt, undermine, and challenge singular temporal framings of crisis. This is not to deny that multiple overlapping crises are currently in effect nor to downplay just how rapidly and rapaciously those effects are destroying the Earth. Rather, it is to maintain a focus on regional realities coupled with a refusal of proposed transition 'solutions' which threaten to exacerbate rather than resolve entrenched socio-environmental inequalities.

From Just Transition to Restorative Environmental Justice

An influential concept in transitions thinking is the 'just transition' which gained visibility from the 1970s onwards through the collective efforts of trade unionists in the US to reconcile environmental and social concerns (while also confronting the employer practice of 'job blackmail'—forcing workers to take on unsafe and toxic jobs or face dismissal) (Stevis et al. 2019). The term now commonly refers to ensuring 'transitional support' such as reemployment, training, reinvestment plans, or benefits for workers and communities who currently depend financially on the fossil fuel industry (Pollin and Callaci 2016). US proposals include adding a 'Just Transition Fee' on the value of oil production to cover lost wages and college costs for oil workers facing job losses, or redirecting state funds to reemploy oil workers in decommissioning projects (e.g. plugging abandoned oil wells) (OCI 2018; Partridge et al. 2020). Improving energy access is another aspect of building a just transition. Global inequalities in the distribution and availability of affordable energy services are stark. Almost four billion people live in 'energy poverty,' without access to electricity or low-pollution modes of cooking, or without the means to sufficiently heat their homes (IEA 2017; Partridge forthcoming-a). Inequalities in access to energy are further marked by spatial, racial, and socioeconomic disparities (Reames 2016). A just transition therefore has to extend access to reliable, clean energy sources for those currently living without while also ensuring that measures introduced to meet those goals do not worsen current environmental inequalities (Newell and Mulvaney 2013).

While just transitions approaches have been used in different ways across activist, intellectual, and advocacy settings to reconcile environmental, social, and climate justice concerns within the global shift toward low-carbon economies and energy systems, dominant policy debates on the topic "have seen a profusion of techno-managerial framings of the process, underpinned by narrow cost-benefit analyses" (Bouzarovski 2022, pp. 1003–1004). Recent critiques highlight how such dominant framings and policy approaches—including the European Green New Deal and a growing body of uncritical scholarly work—pursue a limited concept of justice as a formalized 'thing' to be 'delivered' or 'applied' via linear reform trajectories, with the effect of further obscuring actual disruptive, collective, and emancipatory struggles; perpetuating "new forms of enclosure and division"; and failing to challenge the "capitalist roots of energy and climate injustices" (ibid., citing Velicu and Kaika 2017 on concepts of justice). It is therefore vital that scholarly attempts to draw on the just transition concept, especially as a way to bridge energy and environmental justice concerns, remain connected to and informed by the work of groups who are already applying the concept within on-the-ground actions and projects.

Engaging with ongoing movement-based initiatives reduces the risk that transitions-focused scholarly work repeats past mistakes. Echoing the trajectory of other terms that have been institutionalized and separated from their radical roots—both in the past (e.g. 'sustainability') and currently (e.g. as is arguably the case with 'energy justice'—see Chap. 3, below)—now-dominant perspectives on just transitions overlook the prior work of EJ and other activists who pursue more collaborative, counter-hegemonic approaches. These include UPROSE community resilience activists in Brooklyn who, over decades, have shown how unless just transitions centralize racial justice, then "climate adaption and economic development initiatives are likely to reenact policy violence, specifically by reinforcing market fundamentalist approaches to space and accelerating the displacement of the working class and communities of color" (Sze and Yeampierre 2018, p. 62). Counteracting such processes of displacement in historically Black communities in North Philadelphia by developing local energy initiatives, community leader John Bowie describes the work of Serenity Soular—a solar panel installation company built in collaboration with Swarthmore College faculty and students—in just transition terms with an emphasis on reclaiming community power: "We insist that this business be based in the neighborhood, and owned by its workers, so that we are building, rather than extracting, wealth and knowledge within the community" (Di Chiro and Rigell 2018, p. 93). Many more examples emerge in diverse global contexts where collaborative provisioning efforts seek to address combined threats to local energy, employment, and environmental health needs.

In their multinational analysis of just transition discourse and initiatives, Wilgosh et al. (2022) highlight how labor and environmental movements (including unions and advocacy groups) offer more transformative visions than other actors (such as private sector and governmental agencies) for post-carbon restructuring, in part by rooting key proposals in collective ownership of the commons (Kishimoto et al. 2020) and in methods designed to ensure that energy and wealth are owned and controlled by 'decentralized democratic communities' (Adkin 2017). A first step for analytical work that seeks to better understand the pitfalls and potentials of political contestation around the just transition concept is to understand the movement-based histories of these transformative visions. As members of the Transnational Institute point out, certain EJ groups and members of the trade union movement began building alliances on common ground characterized in the mobilizing slogan 'there are no jobs on a dead planet':

"The movements which are today beginning to consolidate proposals under the name 'just transition' have a rich history. Movements against free trade agreements and neoliberalism; the alter-globalisation movement; energy sovereignty and democracy struggles; environmental justice movements; labour movements; decolonisation and independence struggles; feminist and women's movements; movements against racism; and fights for agrarian reform, peasant rights and food sovereignty, among others, have helped to lay the groundwork for discussions today. This diversity of backgrounds, political traditions, and strategic goals means that the dialogue that is creating a radical concept of just transition is not free from tensions or contradictions. However, it is increasingly clear that a critical mass of organisations or movements see the vital importance of working through these tensions in order to form stronger and more radical alliances for systemic transformation". (TNI 2020, p. 5)

EJ and Indigenous scholars and activists have specifically made the case for an expansive understanding—and application—of the just transition concept. Formulations of 'energy justice' made by Indigenous activists, for example, expand the temporal frame of transitions thinking to also address how histories of exploitation and pollution continue to shape current experiences of injustice and determine potential future energy development (Honor The Earth 2010). Applied to the just transition, these approaches not only address labor and employment concerns, energy access, and climate justice issues, but also a wide range of toxic legacies from energy systems past and present (LaDuke 1999; Razack 2002; Voyles 2015). This means attending to the wellbeing and livelihoods of those who live among the ruins of polluting energy infrastructures and who suffer through the 'afterlife' of destruction caused by prior resource extraction and energy generation (Auyero and Swistun 2009; Gordillo 2014). In support of these ideas, some scholars and practitioners propose 'restorative justice' as a form of reparation for previous systemic injustices, for example through new community redevelopment investments, job training, and housing provision, in addition to environmental remediation and 'clean up' efforts at polluted sites (Dorsey 2009). In certain contexts, particularly within Indigenous struggles for justice, efforts to restore environmental quality and the economic and social vibrancy of affected communities need to go further still: by recognizing treaty rights and the collective governance of non-privatized lands (since disregarding these is frequently a cause of environmental destruction) (Gedicks 2001).

Recognizing the complementarity between defending Indigenous rights and environmental wellbeing, the Indigenous Environmental

Network (IEN) has published a document of twenty-five principles that describe Indigenous values essential to the just transition and to climate justice (Sze 2020). The principles draw on many years of energy-focused activism. IEN, as well as Honor The Earth, have differentiated themselves from mainstream environmental and social justice groups by asserting "pan-Indian identification as the historical difference informing their critique of the 'energy colonialism' experienced by indigenous peoples in the United States and Canada" (Powell 2018, p. 87). The first of the twenty-five Indigenous Principles of Just Transition refers directly to restoration: "A Just Transition affirms the need for restoring indigenous life ways of responsibility and respect to the sacred Creation Principles and Natural Laws of Mother Earth and Father Sky, to live in peace with each other and to ensure harmony with nature, the Circle of Life, and within all Creation" (IEN 2017, p. n).

Other articles from the Indigenous Principles of Just Transition foreground productive relations and work/labor issues:

> "A Just Transition establishes the necessary elements of an economic system consistent with the Rights of Mother Earth to include: Immediately reducing production and consumption levels to within the natural order of Mother Earth; The full restoration of ecosystems, primarily allowing nature to heal itself; Elimination of economic systems and strategies that prioritize economic growth, and profit, and private acquisition of resources and wealth, above all other values; The elimination of substances that are toxic, persistent, and bioaccumulative; Zero waste systems for production, use, and decomposition known as cradle-to-cradle living; Recognition of sacred relationships with place, and; in all economic decisions and human activities, the wellbeing of Mother Earth and the Circle of All Life is primary...

> A Just Transition recognizes that strategies were first forged by labor unions and environmental justice groups, rooted in people of color and low-income communities as well as Indigenous lands; who jointly saw the need to phase out industries that were polluting workers, community and Mother Earth; and at the same time provide just pathways for workers to transition to other jobs. It was rooted in workers defining a transition away from toxic polluting industries in alliances with fence line and frontline communities." (IEN 2017)

Throughout these Principles, acts of solidarity and care across diverse communities are fundamental to the pursuit of transitions-as-justice, including a central role for multiple forms of restoration.

Many peoples and Indigenous communities draw on experiences over generations of developing practices that restore spiritual, emotional, physical, material, and relational wellbeing (Smith 1999; LaDuke 2022). Many activist mobilizations, among them contemporary Indigenous struggles and EJ movements, also adopt an intentional focus on restoring relations of different kinds. A restorative focus might center relations of care with nonhuman kin as way to guide movement actions, commitments, duties, and priorities (Winter 2021). It might also be a way to build and strengthen EJ coalitions (Grossman 2005; Di Chiro 2008). Often, EJ movements combine both. For example #NoDAPL—the movement led by the Standing Rock Sioux and allies to protect water and land from the Dakota Access oil pipeline, focused in 2016 on the Oceti Sakowin camp and treaty lands—was a space where acts of resistance were strengthened through women-led organizing in collaboration with men, gender-nonconforming people, and youth (TallBear 2019). The Red Nation authors clarify that restorative work connects across multiple sites of resistance and is central to what multiple movements are fighting to achieve:

> "[Standing Rock] was part of a constellation of Indigenous-led uprisings across North America and the US-occupied Pacific: Dooda Desert Rock (2006), Unist'ot'en Camp (2010), Keystone XL (2011), Idle No More (2012), Trans Mountain (2013), Enbridge Line 3 (2014), Protect Mauna Kea (2014), Save Oak Flat (2015), Nihígaal Bee Iiná (2015), Bayou Bridge (2017), O'odham Anti-Border Collective (2019), Kumeyaay Defense Against the Wall (2020), and 1492 Land Back Lane (2020), among many more. Each movement rises *against* colonial and corporate extractive projects. But what's often downplayed is the revolutionary potency of what Indigenous resistance stands *for*: caretaking and creating just relations between human and other-than-human worlds on a planet thoroughly devastated by capitalism". (The Red Nation 2021, pp. 16–17, original emphasis)

Building connections between struggles and contexts remains vital to restorative work. Research in care ethics has identified complimentary approaches from Indigenous and feminist philosophies, for example connecting the women-led Chipko movement's application of local knowledge about trees and the many lifeforms they support with Indigenous conceptions of care that attend to reciprocity and the restoration of wounded interspecies relationships (Whyte and Cuomo 2017). Different forms of 'restorative' environmental work therefore seek to strengthen the spaces required to enact these many diverse practices (and to counteract those agencies or individuals who would curtail the freedoms and abilities required to embody those practices). A broader commitment to restorative work and inquiry involves recognizing and supporting the many global, and networked, acts of restoration that are already being pursued.

Important examples abound. Drawing on her work with Afro-Puerto Rican women addressing energy-related injustices, Hilda Lloréns describes collective efforts underway to create restorative alternatives to the processes of dispossession and displacement that are being driven by racialized exploitation, capitalist extraction, and multiple forms of uninvited environmental change (Lloréns 2021). Drawing on work with Afro-Colombian organizations and movements for reparative and transformative justice, Arturo Escobar describes how the concept of "reparation ecologies" emerged from demands for compensation—specifically as a way of identifying "the historical imbalance between the material and cultural contributions of the black communities to nation building and the meager retribution by the nation in terms of investment and conditions for development in areas where black communities predominate" (Escobar 2008, pp. 367–368, lower case original). The concept connects memory, healing, and reconstruction within justice struggles. Reparations here attend to diverse experiences of racism, sexism, and other forms of violence; reparations also aim to empower subaltern subjects (in addition to decolonizing 'hierarchical geographies of race') and to mobilize community members through an equal combination of 'action, reflection, and commitment' (Escobar 2008). More recent work on reparations for slavery specifically reframes the concept as one dimension of a broader, future-oriented project based on new notions of success or progress—looking beyond any one particular reconciliatory or restitution measure and struggling for justice at a 'world scale' in an era of climate change (Táíwò 2022).

The idea of reparations has also been applied specifically within transitions thinking. Patel and Moore (2017) draw on the concept of liberation ecologies—spaces of struggle over power and resources to build emancipatory alternatives—to imagine global, anti-capitalist futures (Watts and Peet 2005). Reparation ecology here is "learning to interact with the web of life differently," a way to simultaneously recognize the violence and inequality of capitalist modernity while also reorganizing aspects of life still rooted in those systems, aiming instead to reconnect humans and ecosystems (Patel and Moore 2017, p. 207). The struggle is to create and keep open spaces for those (re)connections to occur in mutually fulfilling ways—a central goal of restorative environmental justice.

Restorative environmental justice is emergent; a branch of broader EJ thinking and acting that remains open in terms of conceptualization and application. It builds on previous applications of 'restorative justice' that sought to bring together different responses to environmental change, connecting four related issues: environmental quality; social justice; collaborative decision making; and community economic development (Dorsey 2009). A focus in the field of restorative justice on repairing environmental harm has also emphasized the need to include all those concerned in restorative processes—including victims, perpetrators, and the state—but on the condition that justice is defined in light of the claims being made by those who are seeking environmental justice and that guilty (typically corporate) actors are comprehensively held to account (Minguet 2021).

Other precedents to restorative environmental justice include (i) the work of practitioners seeking to connect EJ with traditions and techniques for restorative justice as a form of healing in cases of social conflict and criminal justice (McCaslin 2005); and (ii) connections with ecological restoration, a field of scientific practice defined as "the process of assisting the recovery of an ecosystem that has been degraded, damaged, or destroyed" (SER 2004, p. 3). Ecological restoration covers a wide range of practices. Some adopt a managerial approach focused on restoring ecosystem structure and function, while others are rooted in more holistic methods and community-based work. The latter invites people to "participate in healing the wounds left on the earth, acknowledging the human power to create as well as to destroy" by restoring not only landscapes but also the "diversity-enhancing capabilities of the human communities inhabiting those landscapes" (Nabhan 1991, p. 4). While there are clear parallels here between EJ and ecological restoration, the two areas of practice also diverge in important ways—differences that reinforce the importance of solidarity as a basis for restorative environmental justice.

Practitioners of ecological restoration, as within EJ, have long critiqued approaches that fail to identify and deal with the systemic causes of degradation or that neglect to push for systemic societal change (Hobbs and Norton 1996). But it is diverse Indigenous voices and groups who have led the way for recognizing holistic wellbeing as rooted in the restoration of relationships between all beings. For restoration ecology as a field of practice, applying these insights (or embodying such renewed relationships) means looking beyond concepts of protection to cultural survival and ecological health—while also rejecting dominant framings of 'solutions to sustainability problems' which replicate the same dynamics of exploitation and control that created most of the world's ecological harm (Martinez 2018). Accounts of Indigenous Environmental Justice similarly build on the related concept of "reciprocal restoration," understood as the "mutually reinforcing restoration of land and culture, such that the repair of ecosystem services contributes to cultural revitalization, and renewal of culture promotes restoration of ecological integrity" (Kimmerer 2018, p. 41). These are views that have been historically and systematically marginalized from institutional scientific practice. The notion of knowledge exchange hints at ways of rectifying this dynamic (of power imbalance and inequality between Indigenous persons and persons of other nations and heritages) but it also always risks deepening those differences (Whyte 2018). More reciprocal approaches require, in addition, a conscientious rebuilding of the concepts and practices that shape how coalitions are understood and enacted.

While engaging with concepts such as reciprocal restoration and other insights from diverse Indigenous practices, Western, institutional restoration science has to work against the superficial or exploitative use of Traditional Ecological Knowledge (TEK) or Indigenous Environmental Justice, especially when such misappropriation is used as a way to meet "diversity requirements or to hop on the academic bandwagon" (Nelson 2018, pp. 257–258). Attempts to 'bridge' or 'integrate' TEK and dominant scientific practices operate within historical and entrenched power asymmetries; scientists and others therefore have a responsibility to reject patterns within institutional practices that further silence or harm Indigenous people (Wildcat 2009; Martinez 2010; Whyte 2018). One response is to rethink participation in restoration efforts as a way to relocate restorative work within the lives, livelihoods, and locations of those most affected. Indigenous and critical voices within ecological restoration have rejected the exclusionary effects of technologizing and professionalizing restoration, instead finding ways to support the involvement of as many people as possible in grassroots community action (Nabhan 1991). The goal is respectful alliance-building—another form of solidarity—rather than the amalgamation or appropriation of views held (and practices embodied) by distinct communities or collectives.

Restorative environmental justice, as developed in Julie Sze's recent work, is markedly framed as a way to build solidarity. Attentive to the problematic power dynamics across the field of restoration ecology, restorative environmental justice delivers a statement of intent to work directly against those forms of marginalization and epistemic (and other forms of) violence. Restoring relationships of respect and reciprocity across movements and mobilizations means recognizing the distinct experiences of Indigenous groups, EJ communities, and others engaged in justice-oriented activism while also identifying common opponents and working toward common goals, when appropriate. From these points, Sze builds out restorative environmental justice as a call for solidarity focused on "accountability, art, and the continued search for freedom in a body or bodies shaped by the forces of racism, capitalism, and technology" (Sze 2020, p. 87). This is a vision for a field of thought and action that is "explicitly decolonial and integrative, including humans as animals and imagining humans and nonhuman nature [in] nonextractive modes" (Sze 2020, p. 79). These are ambitious, and often fraught, goals. Nonetheless, the principles of restorative environmental justice are central to the work of reframing transitions as collaborative projects of justice—using transitions as opportunities to radically resist, reorganize, and repair destructive patterns of socioecological inequality.

DEGROWTH

'Degrowth,' as both a concept and a growing social movement, recognizes that the quantity of matter and energy used by human economies (and the exploitation of humans and Nature required to produce those materials) cannot continue to grow without exacerbating multiple planetary crises (Kallis et al. 2020). The era of 'carbon-driven modernity' is also the current era of the Earth's sixth mass extinction event—the increasingly rapid and permanent destruction of global biodiversity—with some calculations plotting extinction rates for terrestrial plants and animals to be "*at least* 1,000 times higher" than the prehuman or 'background' average (Howe 2019, p. 138, original emphasis). At the same time, the rate at which global economic inequality is growing is also increasing, just as labor rights are further eroded and the majority of jobs become increasingly precarious: "it is not only species that are becoming extinct but also the words, phrases, and gestures of human solidarity" (Guattari 2000, p. 44). On a basis of collective action, ecological economics, and EJ analyses, degrowth builds a holistic response to these inter-related processes of devaluing and destruction. The degrowth movement calls for a planned and equitable reduction—and radical transformation—of the global economy, with redistribution of wealth and societal shifts toward shared values of care, the commons, and solidarity (Kallis et al. 2015).

As non-renewable resources are being rapidly depleted globally (and the rate of consumption for some critical renewable resources such as fish and timber outpaces the rate of recovery), there are calls for "dematerialization" to reduce social metabolism and to decrease overall global flows of energy and materials (Lorek 2015). These concerns are articulated across a wide range of environmental groups and actions. Text in the recent IPCC report (April 2022) reaffirms that economic growth—and specifically growing GDP per capita—remains "by far the strongest upward driver" of global greenhouse gas emissions (IPCC 2022 [AR6 WGIII, Section 2.4.1, subject to copyedit]). Underlining how total net emissions continued to rise during 2010–2019, the same report confirmed that only rapid and radical changes to global energy systems and economic activity will prevent our world from becoming unlivable (ibid.). As a field of study and a global movement for transformative change, though, degrowth has goals that look beyond the potential of technological innovation or energy efficiency to reduce global metabolic flows.

Foundational texts for degrowth and ecological economics scholarship include the works of Georgescu-Roegen, among them the (1975) article "Energy and Economic Myths" in which he details a series of recommendations for reducing global social metabolism and minimizing societies' overexploitation of the planet. [3] The very first proposal reads: "the production of all instruments of war, not only of war itself, should be prohibited completely" (p. 377). It is an always-relevant reminder that addressing economic growth and inequality requires action across multiple social, political, temporal, and geographical scales. At the time of writing (June 2022), as the immeasurable human suffering and ecological toll of Russian president Putin's invasion of Ukraine continue to worsen, the importance of demilitarizing human economies as a way of degrowing human economies is, yet again, being acutely and painfully made clear.

---

[3] Thanks to Erik Gómez Baggethun and members of the Ecological Economics and Degrowth reading groups at ICTA-Barcelona for highlighting the immediate, contemporary (and ongoing) relevance of this particular article.

Degrowth approaches consider the concentric spheres of industrial activity that are required to fuel economic growth. Environmental injustices can occur at any point along complex, globalized chains of production and across the life cycle of different technologies (Mulvaney 2019; Levenda et al. 2021). Environmental injustices also occur along transportation (and transmission) routes of resources, minerals, and energy itself (Coyotecatl Contreras 2016; Banschbach and Rich 2021). At the same time, unless systems for the provision of other basic needs along with energy—especially water and food—are designed specifically such that the input requirements and waste outputs of one do not adversely affect another, then access to these basic needs is further jeopardized and the likelihood of environmental injustices occurring is increased. This is particularly true in contexts where people and places are subject to historical and contemporary processes of neo-colonialism and impoverishment (Verhoeven 2021). Degrowth perspectives therefore extend the critical focus of justice-related energy research far beyond any one particular energy initiative or energy source.

Degrowth stands in opposition to 'eco-modernist' visions which imagine technological applications will allow societies to avoid the environmental crises of capitalism—visions which also ignore how contemporary energy practices depend on the exploitation and destruction of marginalized peoples and places and on the underdevelopment of the Global South (Bell et al. 2020, p. 6). This touches on a point of overlap between EJ and degrowth: both recognize that an equitable distribution of the benefits and externalities of energy systems (as well as the production/consumption of energy across different scales) requires structural transformations of productive systems. Some EJ work claims energy transitions require a shift 'from resistance to reconstruction' and to a politics of 'sustainable materialism' in our everyday material relationships (Schlosberg 2013). Degrowth, however, is here aligned more closely with the work of ecofeminist writers who point out that a focus on the politics of everyday materiality is not new: "ecofeminist theory *has never not* been grounded in materiality… [it] is built on revaluing provisioning work, foregrounding embodied difference, and making the personal political" (MacGregor 2021, p. 42, original emphasis). Thus, despite differences between EJ and degrowth in terms of movement histories and strategies, the two fields connect through (i) a rejection of extractivism and economies that operate without regard for social and environmental costs (Rodríguez-Labajos et al. 2019); and (ii) a combined focus on materiality, relationality, and resource use.

EJ and degrowth also share a recognition of how experiences of environmental injustice are produced: not as isolated incidents but as the result of particular ways of organizing and coordinating production, mechanization, and economic growth. Energy is the pivotal factor linking each of these processes. In terms of physics, energy can be described as a fundamental dimension of existence that dynamizes and shapes our living experience. But 'energy' as it is more commonly understood today—in terms of work and power—is a concept that emerged within a particular sociomaterial context.

What became recognized as an energy 'sector' emerged amid "the convergence of bodies, fossil fuels, and steam engines in imperial Europe and its factories" that also generated, and subsequently reinforced, a dominant industrial perspective (Daggett 2019, p. 6). Newton's laws of mass and motion affirmed Descartes' view of the universe as a 'vast machine' and framed the modern European worldview: a 'scientific materialism' that studied life according to mathematical rules, sought to free society from superstitions, and provided the technical foundation for industrial society—in the process entrenching the myth of human dominance (Rees 2019, pp. 134–135). This nineteenth-century scientific materialism—what Gaston Bachelard has described as a "dematerialized materialism"—was predicated on energy as a pervasive, invisible force behind all motion and matter: "the materiality of the physical universe—energy—was nowhere to be encountered except in the manifest consequences of its enormous labor power" (Rabinbach 1992, pp. 48–49). The application of universal laws of energy to the workings of the cosmos was mirrored by the application of 'natural laws' of development to the workings of society (favoring productivity, performance, and progress) and a reconceptualization of the human body as a 'working machine' (Rabinbach 1992, pp. 49; 51). The industrial convergence of bodies, fuels, and machines became the basis for dominant understandings of progress, development, and growth that we—globally—still live with today.

Combined with a shift in science toward energetics (the conservation and exchange of energy), scientific materialism supported the systematic study of social-ecological metabolism (Stoffwechsel); for Marx, building on the prior work of Liebig, Mayer, Moleschott and others, this referred to the interaction of society and nature through labor (Foster 2000, p. 160). At the end of the nineteenth century, metabolic and biophysical perspectives on economics were not uncommon, but only in the 1980s were schools of thought established (among them ecological economics and industrial ecology) that specifically used social metabolism to analyze the economy within a physical system described in terms of flows of energy and materials including water (Martinez-Alier 1987, 2009). These models distinguish between energy use as endosomatic (within the body, as food) or exosomatic (as fuel for cooking/heating or as power to operate machines/appliances) and analyze (i) the increasing social metabolism and unequal exosomatic energy use associated with economic growth and (ii) the increasing environmental conflicts that result (Martinez-Alier 2009, pp. 62–63). These models further make use of the concepts of "net energy" or EROI ("energy return on investment") to evaluate different fuel sources by calculating the ratio between the energy delivered to society by a particular fuel and the energy spent in harnessing and delivering that energy (Hall et al. 2014). By engaging directly with the materiality of energy—and the material consequences of expanding energy use—socio-metabolic analyses offer tools for scrutinizing energy sources and systems (and related injustices) at multiple different scales.

These insights also offer critical input for transitions thinking. The macro perspective of social metabolism orientates 'transitions' as moments of systemic shift, urging industrial resource and energy use to 'transition' into realignment with the adaptive and regenerative capacities of the biosphere (Fischer-Kowalski and Haberl 2007). Quantifying systemic flows using metabolic models can also cast new light on systemic flaws in contemporary energy-society relations. For example, using recent national data for Spain and regional data for Barcelona, researchers found that (i) annual energy *inputs* for agricultural production in Spain are six times higher than the energy *content* of all the food produced (Infante Amate and González de Molina 2013; Sekulova et al. 2013), and (ii) service delivery in the commercial sector (for example, privatized caring services) is significantly more energy intensive than equivalent activities within households (D'Alisa and Cattaneo 2013). These studies clearly illustrate the need for systemic alternatives in the agricultural and healthcare sectors—not only concerning the overall amounts of energy being used but also how decisions are being made (and who is being excluded). When macro-scale analyses calculate or characterize national level averages of energy use and impacts, those averages risk obscuring variations in how those inequalities are experienced (Torres and Marques 2001; Hess and Ribeiro 2016). For this reason—seeking to directly address such inequalities and variations—degrowth maintains a focus on how power asymmetries across society(s) and processes of marginalization underpin energy- and resource-intensive modes of economic activity.

Other macro-scale perspectives on energy-society relations focus primarily on issues of energy access and the notion of 'energy sufficiency.' This work recognizes undeniable, instrumental links between more intensive energy flows and increased environmental degradation while also identifying ambivalence in the relations between energy use and human wellbeing (Smil 2004; Chapman 2013). In analyzing these relations, one proposed branch of energy ethics (type I) assumes linear links between levels of energy consumption and human well-being, therefore limiting the analytical focus to questions of ethical energy production (Mitcham and Rolston 2013; Geerts 2017). Another approach (type II energy ethics) instead tackles the conundrum directly: acknowledging that access to energy can support health and well-being while also recognizing that assured ecological damage results from net increases in energy consumption.

In type II energy ethics, energy is seen as a qualified good (something that is good only under certain circumstances, its benefits conditional on those circumstances): energy consumption is seen as supportive of human flourishing, but only up to a certain point, beyond which its societal and ecological impacts become damaging and counterproductive (Mitcham and Rolston 2013, p. 318). That point, according to this model of energy ethics, can be plotted on a saturation curve where the notion of human development decouples from energy consumption and there is effectively no rise in the former relative to the latter for populations who already consume large amounts of energy (Martínez and Ebenhack 2008). The 'energy sufficiency' concept is used to study such thresholds and to establish maximums and minimums of energy consumption, designing energy systems primarily around the fulfilment of basic human needs (food, health/hygiene, lighting, adequate heating/cooling, and information/communication) rather than technical specifications or (only) emissions targets (Thomas et al. 2015). Current levels of atmospheric CO_2 are largely the result of consumption patterns in countries where most of the world's wealth is accumulated and concentrated. Recognizing this, some energy sufficiency scholars focus on "targeted reductions and qualitative changes in energy use among high-energy communities and societies" (Burke 2020, p. 13). The kinds of energy transition being proposed therefore specifically aim to reduce excessive consumption.

Definitions of sufficientarianism, however, are inevitably context-driven and context-dependent (Monyei et al. 2018; Millward-Hopkins et al. 2020). As such, and as with other technocratic approaches to energy and environmental crises, dominant perspectives from places with high current levels of energy use continue to have an outsize influence on how notions of sufficiency are defined (and how different needs or values are understood). Degrowth takes a different approach. It relates energy use to redistribution, to questions of power and autonomy, and seeks a more expansive platform for linking energy provisioning to global, social, and environmental (in)equality. Gerber et al. emphasize that degrowth not only addresses the material conditions of (re)production and the quantification of metabolic flows (by asking if those flows are excessive or sufficient); it also scrutinizes the power relations that shape those metabolic flows (by asking who designs and controls the flows of energy and resources through societies) (Gerber et al. 2020). It is therefore a political project that, like many EJ movements, creates and engages with different languages of valuation that extend beyond material and economic concerns to also include ethical, aesthetic, spiritual, and cultural considerations (Martinez-Alier 2002). By taking these plural values seriously, projects of degrowth are also projects that decenter dominant definitions of progress, sufficiency, development, and 'the good.'

Degrowth also partially reflects a reclaiming of non-mechanized understandings of energy based in a very different politics of labor—focused not only on the material potential for action but also on the dynamism and possibilities of being-in-the-world (Daggett 2019). The importance of conviviality as a shared value within degrowth projects reflects this reclamation of energy as a form of human potential and interconnection, as in Ivan Illich's vision of a convivial society: a "society-wide inversion of present industrial consciousness" such that we "relearn to depend on each other" rather than on ever-expanding energy infrastructures (Illich 1973, p. 27). Through such processes of unlearning, relearning, reclamation, and reorganization, diverse degrowth approaches promote shared values based on the commons and mutuality, further connecting with multiple movements for new energy initiatives.

Energy Sovereignty

There is significant overlap between degrowth objectives and commons-based approaches to resource management: counteracting dominant perspectives that assume relations of (inter)dependence are based primarily on hierarchy and competition (Kenrick 2009) and re-commoning control over the means of production, consumption, trade, and reproduction (Kothari 2019). Degrowth and EJ movements outline critical, historical perspectives on the effects of different modes of production, foregrounding collective efforts to counteract an antidemocratic concentration of resources and political power through localizing control of energy infrastructures and decision-making processes. Energy decentralization is one of a number of EJ narratives that contribute to ideas and practices for degrowth. Around wind power these also include the defense of 'energy sovereignty' (Mexico), 'territorial autonomy' (Mexico and Western Sahara), 'limits to economic growth' (Greece), and the promotion of wind power cooperatives (Avila 2018). These are projects that place the material impacts of energy practices in the broader context of shared values and visions for society(s) as a whole.

Collective ownership (of energy infrastructure) offers at least the potential for more directly accountable decision-making processes, reduced per capita energy consumption, and the integration of ecological principles into energy businesses, in contrast to private or corporate models (Kunze and Becker 2015). Collective ownership may also support broader input into the design of energy systems—for example with regard to questions of durability, reparability, and the purposes of energy use—a responsiveness that is incompatible with typical large-scale, growth-oriented energy initiatives (Kerschner et al. 2018). Many different examples of 'community energy' projects take shape globally across diverse political, social, and geographical contexts, where communities of place or interest increase their control or ownership of energy infrastructure and where the benefits of either supply- or demand-side sustainable energy initiatives are shared collectively (Seyfang et al. 2013). Such initiatives are context-dependent, emergent, and responsive to the complex mix of values and processes that allow communities to function (Creamer et al. 2019).

Community energy initiatives consequently cover an incredible range, from localized renewable energy systems and community hall refurbishments in the UK (where many civil society-led groups remain dependent on outside sources of support) (ibid.), to proactive efforts in Taiwan to redefine both socioecological relations and political, economic, and material relations with outside institutions (including reducing dependency on the fossil fuel industry), specifically through proactive, place-based environmental justice activism (Lai 2022). Other cases, for example in Ontario, Canada, illustrate just how deep the 'incumbency' often is of traditional economic, energy, and political structures, constraining any positive impacts of community energy initiatives (Brisbois 2019). Further cases and experiences, conversely, illustrate the potential of community-based energy initiatives to radically redirect regional politics toward emancipatory goals, as seen in efforts in Berlin and Barcelona (Angel 2017; Ortiz et al. 2021). These and many other cases show how the reconfiguration of energy systems—when aiming to improve local environmental relations, reduce climate change impacts, and when led primarily by citizens and others outside of established seats of (government and corporate) power—can influence and reshape political and energy systems at the national level and beyond.

As with natural resources in general, energy is inherently political not only because it is an object of conventional politics but also because it is materially embedded in the socio-technical formations through which political processes unfold (Huber 2019). Thus collective or community ownership is a vehicle both for localizing energy and income generation *and* for redefining or redemocratizing political action—as when the 'community' in Community Renewable Energy initiatives becomes a platform for solidarity and more equitable, and practical, political participation (Kumar and Taylor Aiken 2021). Very often, the inverse is also true: *prior* histories of community organizing and cooperation (together with more directly accountable models of land ownership) are very significant in determining, or guiding, the relative success of any community-owned/operated energy initiative.

See the APPENDIX for information on two different models of collective ownership and participation that have informed the design, use, and governance of localized energy initiatives. One is located in Gujarat state in India, the other on the small island of Eigg off the west coast of Scotland. Despite very different social, political, economic, and geographical contexts, the two cases both offer insights for how the restructuring of energy relations connect with other social issues and transformative goals.

Projects of energy localization offer another way to repoliticize energy relations, to re-examine relations of dependence and production, and therefore also to chart further potential routes toward degrowth objectives. Central here is the concept of 'energy sovereignty' which has emerged from global social movements, drawing on a range of EJ struggles and insights (Del Bene et al. 2019; Angel 2020). Energy sovereignty places decentralization, inclusive and directly democratic decision-making processes, and respect for the regeneration of ecological cycles at the heart of redesigning and relocalizing energy systems (XSE 2018; Del Bene et al. 2019). Reflecting movements for political and food sovereignty, it is rooted in control over territory and productive resources and maintains that localized producers should determine what is produced and how, under what conditions, and for what purposes (Partridge 2020, forthcoming-a). Movements for energy sovereignty connect closely with parallel struggles for access to food and food justice.

The diverse global movement, La Vía Campesina, has been a leading voice in articulating and fighting for food sovereignty as part of ongoing struggles against industrialized, resource-intensive, and deregulated models of food production and in support of more holistic approaches (Desmarais 2002). Food sovereignty proposes increased direct democratic control over systems of food production by those who are most directly involved. This is a political reorientation of decision-making processes that also explicitly addresses gender rights and builds on anti-discrimination praxis (Vía Campesina 1996). Reflected in both food sovereignty and energy sovereignty principles are tools for reshaping relationships between producers and consumers, fostering not only clearer lines of accountability and transparency but also greater awareness of connectedness and interdependence.

The food sovereignty movement identifies *root causes* of hunger (in terms of energy research the equivalent here would be 'energy poverty'): a lack of food (or lack of access to energy) reflects a lack of socio-economic and political power, situations that are the outcome of antidemocratic concentrations of wealth and power among elites (Lappé et al. 1998; Trauger 2014). Global social movements have mobilized to redress such power asymmetries, for example by: reclaiming control over exchange relations and the markets that govern access (to food or to energy); restructuring systems of production such that producers themselves decide how to use their land and labor; and allowing producers to have a say in how their products will be used (Partridge 2012, 2016, forthcoming-a).

Food sovereignty also aims to address systemic inequalities by ensuring people have a direct political stake in the (food/energy) systems on which they depend (Heynen et al. 2012). These moves are directly reflected in principles of energy sovereignty: the "decentralization, relocalization, and differentiation of energy generation, technology, and knowledge" controlled by both urban and rural communities (Del Bene et al. 2019, p. 178). As noted above (in Chap. 1: Settler Colonialism and Sovereignty) these principles take on very different meanings in contexts where Native and Indigenous sovereignties are being defended—where Indigenous sovereignty is always contained and constricted by settler sovereignty (Powell 2018; Yazzie 2018). Fundamentally, energy sovereignty principles connect closely with the goals of transitions as collaborative projects of justice: using energy localization as a way to counteract the influence of corporate power and instead to (re)localize power within communities and collectives.

REFERENCES

Adkin, L. 2017. Crossroads in Alberta: Climate Capitalism or Ecological Democracy. *Socialist Studies/Études Socialistes* 12 (1): 2–31.

Angel, J. 2017. Towards an Energy Politics In-Against-and-Beyond the State: Berlin's Struggle for Energy Democracy. *Antipode* 49 (3): 557–576.

———. 2020. New Municipalism and the State: Remunicipalising Energy in Barcelona, from Prosaics to Process. *Antipode* 53 (2): 524–545.

Auyero, J., and D.A. Swistun. 2009. *Flammable: Environmental Suffering in an Argentine Shantytown*. Oxford University Press.

Avila, S. 2018. Environmental Justice and the Expanding Geography of Wind Power Conflicts. *Sustainability Science* 13 (3): 599–616.

Awâsis, S. 2020. 'Anishinaabe time': Temporalities and Impact Assessment in Pipeline Reviews. *Journal of Political Ecology* 27 (1): 830–852.

Banschbach, V., and J.L. Rich, eds. 2021. *Pipeline Pedagogy: Teaching About Energy and Environmental Justice Contestations*. Cham: Springer.

Bargh, M. 2021. Diverse Indigenous Environmental Identities: Māori Resource Management Innovations. In *Routledge Handbook of Critical Indigenous Studies*, ed. B. Hokowhitu, A. Moreton-Robinson, L. Tuhiwai-Smith, C. Andersen, and S. Larkin, 420–430. New York: Routledge.

Bell, S.E., C. Daggett, and C. Labuski. 2020. Toward Feminist Energy Systems: Why Adding Women and Solar Panels Is Not Enough. *Energy Research & Social Science* 68: 101557.

Bouzarovski, S. 2022. Just Transitions: A Political Ecology Critique. *Antipode* 54 (4): 1003–1020.

Brisbois, M.C. 2019. Powershifts: A Framework for Assessing the Growing Impact of Decentralized Ownership of Energy Transitions on Political Decision-Making. *Energy Research & Social Science* 50: 151–161.

Brugge, D., T. Benally, and E. Yazzie-Lewis. 2006. So a Lot of the Navajo Ladies Became Widows. In *The Navajo People and Uranium Mining*, xv–xix. Albuquerque: University of New Mexico Press.

Burke, M.J. 2020. Energy-Sufficiency for a Just Transition: A Systematic Review. *Energies* 13 (10): 2444.

Chapman, C. 2013. Multinatural Resources: Ontologies of Energy and the Politics of Inevitability in Alaska. In *Cultures of Energy: Power, Practices, Technologies*, ed. S. Strauss, S. Rupp, and T. Love, 96–109. Walnut Creek: Left Coast Press.

Coleman, L. 2021. Afterword: People Thinking Energetically. In *Ethnographies of Power: A Political Anthropology of Energy*, ed. T. Loloum, S. Abram, and N. Ortar, 180–194. New York: Berghahn Books.

Coyotecatl Contreras, J.M. 2016. Los espacios de transportación en la economía extractivista. El caso del gasoducto Morelos en el centro de México. *Ecología Política* 51: 108–112.

Creamer, E., G. Taylor Aiken, B. van Veelen, G. Walker, and P. Devine-Wright. 2019. Community Renewable Energy: What Does It Do? Walker and Devine-Wright (2008) Ten Years on. *Energy Research & Social Science* 57: 101223.

D'Alisa, G., and C. Cattaneo. 2013. Household Work and Energy Consumption: A Degrowth Perspective. Catalonia's Case Study. *Journal of Cleaner Production* 38: 71–79.

Daggett, C.N. 2019. *The Birth of Energy: Fossil Fuels, Thermodynamics, and the Politics of Work*. Durham: Duke University Press.

Del Bene, D., J.P. Soler, and T. Roa. 2019. Energy Sovereignty. In *Pluriverse: A Post-development Dictionary*, ed. A. Kothari, A. Salleh, A. Escobar, F. Demaria, and A. Acosta, 178–181. New Delhi: Tulika Books.

Desmarais, A. 2002. Peasants Speak – The Vía Campesina: Consolidating an International Peasant and Farm Movement. *The Journal of Peasant Studies* 29 (2): 91–124.

Di Chiro, G. 2008. Living Environmentalisms: Coalition Politics, Social Reproduction, and Environmental Justice. *Environmental Politics* 17 (2): 276–298.

Di Chiro, G., and L. Rigell. 2018. Situating Sustainability Against Displacement: Building Campus-Community Collaboratives for Environmental Justice from the Ground Up. In *Sustainability: approaches to environmental justice and social power*, ed. J. Sze, 76–101. New York University Press.

Dillon, G., ed. 2012. *Walking the Clouds: An Anthology of Indigenous Science Fiction*. Tucson: University of Arizona Press.

Dorsey, J.W., 2009. Restorative Environmental Justice: Assessing Brownfield Initiatives, Revitalization, and Community Economic Development in St. Petersburg, Florida. *Environmental Justice,* 2 (2), 69–78.

Escobar, A. 2008. *Territories of Difference: Place, Movements, Life, Redes.* Durham: Duke University Press.

———. 2015. Degrowth, Postdevelopment, and Transitions: A Preliminary Conversation. *Sustainability Science* 10 (3): 451–462.

Esteva, G. 2005. Celebration of Zapatismo. *Humboldt Journal of Social Relations,* 29 (1: Zapatismo as Political and Cultural Practice), 126–167.

Fischer-Kowalski, M., and H. Haberl, eds. 2007. *Socioecological Transitions and Global Change: Trajectories of Social Metabolism and Land Use.* Cheltenham: Edward Elgar.

Foster, J.B. 2000. *Marx's Ecology: Materialism and Nature.* New York: Monthly Review Press.

Gandhi, M. 1909. *Hind Swaraj Or Indian Home Rule [1938].* Ahmedabad: Navajivan.

Gedicks, A. 2001. *Resource Rebels: Native Challenges to Mining and Oil Corporations.* Cambridge: South End Press.

Geerts, R.-J. 2017. Towards a Qualitative Assessment of Energy Practices: Illich and Borgmann on Energy in Society. *Philosophy & Technology* 30 (4): 521–540.

Georgescu-Roegen, N. 1975. Energy and Economic Myths. *Southern Economic Journal* 41 (3): 347–381.

Gerber, J., B. Akbulut, F. Demaria, and J. Martínez-Alier. 2020. Degrowth and Environmental Justice: An Alliance Between Two Movements? In *Environmental Justice: Key Issues,* ed. B. Coolsaet, 94–106. Oxford: Routledge.

van Gerven, J.P. 2022. *The Anti-Nuclear Power Movement and Discourses of Energy Justice.* Lanham: Lexington Books.

Gibson-Graham, J.K., and CEC. 2017. *Cultivating Community Economies: Tools for Building a Liveable World.* Community Economies Collective (CEC) / Next Systems Project.

Gordillo, G. 2014. *Rubble: The Afterlife of Destruction.* Durham: Duke University Press.

Gross, L.W. 2014. *Anishinaabe Ways of Knowing and Being.* Farnham: Ashgate.

Grossman, Z. 2005. Unlikely Alliances: Treaty Conflicts and Environmental Cooperation Between Native American and Rural White Communities. *American Indian Culture and Research Journal* 29 (4): 21–43.

Guattari, F. 2000. *The Three Ecologies.* London: Athlone Press.

Gutierrez, G.M., D.E. Powell, and T.L. Pendergrast. 2021. The Double Force of Vulnerability: Ethnography and Environmental Justice. *Environment and Society* 12 (1): 66–86.

Hall, C.A.S., J.G. Lambert, and S.B. Balogh. 2014. EROI of Different Fuels and the Implications for Society. *Energy Policy* 64: 141–152.

Healy, N., and J. Barry. 2017. Politicizing Energy Justice and Energy System Transitions: Fossil Fuel Divestment and a "just transition". *Energy Policy* 108: 451–459.

Hess, C.E.E., and W.C. Ribeiro. 2016. Energy and Environmental Justice: Closing the Gap. *Environmental Justice* 9 (5): 153–158.

Heynen, N., H.E. Kurtz, and A. Trauger. 2012. Food Justice, Hunger and the City: Food, Hunger and the City. *Geography Compass* 6 (5): 304–311.

Hickel, J. 2019. Degrowth: A Theory of Radical Abundance. *Real-World Economics Review* 87: 54–68.

Hobbs, R.J., and D.A. Norton. 1996. Towards a Conceptual Framework for Restoration Ecology. *Restoration Ecology* 4 (2): 93–110.

Honor The Earth. 2010. *Annual Report: Solar Eclipses Coal on the Navajo Reservation*. Minneapolis: Honor The Earth.

Hopkins, R. 2008. *The Transition Handbook: From Oil Dependency to Local Resilience*. Totnes: Green Books.

———. 2019. Transition Movement. In *Pluriverse: A Post-Development Dictionary*, ed. A. Kothari, A. Salleh, A. Escobar, F. Demaria, and A. Acosta, 317–320. New Delhi: Tulika Books.

Howe, C. 2019. *Ecologics: Wind and Power in the Anthropocene*. Durham: Duke University Press.

Howe, C., and D. Boyer. 2016. Aeolian Extractivism and Community Wind in Southern Mexico. *Public Culture* 28 (2 79): 215–235.

Huber, M. 2019. Resource Geography II: What Makes Resources Political? *Progress in Human Geography* 43 (3): 553–564.

IEA. 2017. *Energy Access Outlook: From Poverty to Prosperity*. Paris: International Energy Agency.

IEN. 2017. *Indigenous Principles of Just Transition*. Bemidji, MN: Indigenous Environmental Network.

Illich, I. 1973. *Tools for Conviviality*. New York: Fontana.

Infante Amate, J., and M. González de Molina. 2013. 'Sustainable de-growth' in Agriculture and Food: An Agro-ecological Perspective on Spain's Agri-food System (Year 2000). *Journal of Cleaner Production* 38: 27–35.

IPCC. 2022. *Climate Change 2022: Mitigation of Climate Change (Working Group III contribution to the WGIII Sixth Assessment Report of the Intergovernmental Panel on Climate Change)*. UN: IPCC.

Jaffee, L., and K. John. 2018. Disabling Bodies of/and Land: Reframing Disability Justice in Conversation with Indigenous Theory and Activism. *Disability and the Global South* 5 (2): 1407–1429.

Johnson, V.A. 2017. Bringing Together Feminist Disability Studies and Environmental Justice. In *Disability Studies and the Environmental Humanities: Toward an Eco-Crip Theory*, ed. S.J. Ray and J. Sibara, 73–93. Lincoln: University of Nebraska Press.

Jones, C. 2014. *Routes of Power: Energy and Modern America*. Harvard University Press.

Kallis, G., F. Demaria, and G. D'Alisa. 2015. Introduction: Degrowth. In *Degrowth: A Vocabulary for a New Era*, ed. G. D'Alisa, F. Demaria, and G. Kallis. New York: Routledge.

Kallis, G., S. Paulson, G. D'Alisa, and F. Demaria. 2020. *The Case for Degrowth*. Cambridge: Polity Press.

Kenrick, J. 2009. Commons Thinking: The Ability to Envisage and Enable a Viable Future Through Connected Action. In *The Handbook of Sustainability Literacy: Skills for a Changing World*, ed. A. Stibbe, 33–38. Totnes: UIT Cambridge.

Kerschner, C., P. Wächter, L. Nierling, and M.-H. Ehlers. 2018. Degrowth and Technology: Towards Feasible, Viable, Appropriate and Convivial Imaginaries. *Journal of Cleaner Production* 197: 1619–1636.

Kimmerer, R.W. 2018. Mishkos Kenomagwen, the Lessons of Grass: Restoring Reciprocity with the Good Green Earth. In *Traditional Ecological Knowledge: Learning from Indigenous Practices for Environmental Sustainability*, ed. M.K. Nelson and D. Shilling, 27–56. Cambridge University Press.

Kishimoto, S., L. Steinfort, and O. Petitjean, eds. 2020. *The Future Is Public: Towards Democratic Ownership of Services*. Amsterdam: Transnational Institute.

Kothari, A. 2014. Radical Ecological Democracy: A Path Forward for India and Beyond. *Development* 57 (1): 36–45.

———. 2019. Radical Well-Being Alternatives to Development. In *Research Handbook on Law, Environment and the Global South*, ed. P. Cullet and S. Koonan, 64–85. Cheltenham: Edward Elgar.

Kothari, A., A. Salleh, A. Escobar, F. Demaria, and A. Acosta, eds. 2019. *Pluriverse: A Post-development Dictionary*. New Delhi: Tulika Books and Authorsupfront.

Kumar, A., and G. Taylor Aiken. 2021. A Postcolonial Critique of Community Energy: Searching for Community as Solidarity in India and Scotland. *Antipode* 53 (1): 200–221.

Kunze, C., and S. Becker. 2015. Collective Ownership in Renewable Energy and Opportunities for Sustainable Degrowth. *Sustainability Science* 10 (3): 425–437.

LaDuke, W. 1999. *All Our Relations: Native Struggles for Land and Life*. Cambridge; Minneapolis: South End Press/Honor the Earth.

———. 2022. The Seventh Fire and the Sitting Bull Plan: An Indigenous Green New Deal. In *Routledge Handbook on the Green New Deal*, ed. K. Tienhaara and J. Robinson. London: Routledge.

Lai, H.-L. 2022. Foregrounding the Community: Geo-historical Entanglements of Community Energy, Environmental Justice, and Place in Taihsi Village, Taiwan. *Environment and Planning E: Nature and Space* 5 (2): 666–693.

Lappé, F., J. Collins, P. Rosset, and L. Esparza. 1998. *World Hunger: Twelve Myths*. New York: Grove Press/Food First.

Lennon, M. 2017. Decolonizing Energy: Black Lives Matter and Technoscientific Expertise Amid Solar Transitions. *Energy Research & Social Science* 30: 18–27.

Levenda, A.M., I. Behrsin, and F. Disano. 2021. Renewable Energy for Whom? A Global Systematic Review of the Environmental Justice Implications of Renewable Energy Technologies. *Energy Research & Social Science* 71: 101837.

Lloréns, H. 2021. *Making Livable Worlds: Afro-Puerto Rican Women Building Environmental Justice.* Seattle: University of Washington Press.

Lorek, S. 2015. Dematerialization. In *Degrowth: A Vocabulary for a New Era,* ed. G. D'Alisa, F. Demaria, and G. Kallis, 110–112. New York: Routledge.

MacGregor, S. 2021. Making Matter Great Again? Ecofeminism, New Materialism and the Everyday Turn in Environmental Politics. *Environmental Politics* 30 (1–2): 41–60.

Maguire, P. 1987. *Doing Participatory Research: A Feminist Approach.* Amherst: The Centre for International Education, University of Massachusetts, No. Paper 1.

Malm, A. 2016. *Fossil Capital: The Rise of Steam Power and the Roots of Global Warming.* London: Verso.

———. 2021. *How to Blow Up a Pipeline: Learning to Fight in a World on Fire.* London: Verso.

Martinez, D. 2010. The Value of Indigenous Ways of Knowing to Western Science and Environmental Sustainability. *Journal of Sustainability Education,* May 9.

———. 2018. Redefining Sustainability through Kincentric Ecology: Reclaiming Indigenous Lands, Knowledge, and Ethics. In *Traditional Ecological Knowledge: Learning from Indigenous Practices for Environmental Sustainability,* ed. M.K. Nelson and D. Shilling, 139–174. Cambridge University Press.

Martínez, D.M., and B.W. Ebenhack. 2008. Understanding the Role of Energy Consumption in Human Development Through the Use of Saturation Phenomena. *Energy Policy* 36 (4): 1430–1435.

Martinez-Alier, J. 1987. *Ecological Economics: Energy, Environment and Society.* Oxford: Blackwell.

———. 2002. *The Environmentalism of the Poor: A Study of Ecological Conflicts and Valuation.* Cheltenham: Edward Elgar.

———. 2009. Social Metabolism, Ecological Distribution Conflicts, and Languages of Valuation. *Capitalism Nature Socialism* 20 (1): 58–87.

McCaslin, W.D., ed. 2005. *Justice as Healing: Indigenous Ways.* St Paul: Living Justice Press.

Melamed, J. 2011. *Represent and Destroy: Rationalizing Violence in the New Racial Capitalism.* University of Minnesota Press.

Millward-Hopkins, J., J.K. Steinberger, N.D. Rao, and Y. Oswald. 2020. Providing Decent Living with Minimum Energy: A Global Scenario. *Global Environmental Change* 65: 102168.

Minguet, A. 2021. Environmental Justice Movements and Restorative Justice. *The International Journal of Restorative Justice* 4 (1): 60–80.

Mitcham, C., and J.S. Rolston. 2013. Energy Constraints. *Science and Engineering Ethics* 19 (2): 313–319.

Monyei, C.G., A.O. Adewumi, and K.E.H. Jenkins. 2018. Energy (in)justice in Off-grid Rural Electrification Policy: South Africa in Focus. *Energy Research & Social Science* 44: 152–171.

Mulvaney, D. 2019. *Solar Power: Innovation, Sustainability, and Environmental Justice*. Oakland: University of California Press.

Nabhan, G.P. 1991. Restoring and Re-storying the Landscape. *Ecological Restoration* 9 (1): 3–4.

Nelson, M.K. 2018. Conclusion: Back in Our Tracks – Embodying Kinship as If the Future Mattered. In *Traditional Ecological Knowledge: Learning from Indigenous Practices for Environmental Sustainability*, ed. M.K. Nelson and D. Shilling, 250–266. Cambridge University Press.

Newell, P., and D. Mulvaney. 2013. The Political Economy of the 'just transition'. *The Geographical Journal* 179 (2): 132–140.

OCI. 2018. *The Sky's Limit California: Why the Paris Climate Goals Demand that California Lead in a Managed Decline of Oil Extraction*. Washington, DC: Oil Change International.

Ordoñez Muñoz, S. 2021. Mining and Green New Deals. *The Ecologist*, August 4.

Ortiz, J., M. Jiménez Martínez, A. Alegría-Sala, S. Tirado-Herrero, I. González Pijuan, M. Guiteras Blaya, and L. Canals Casals. 2021. Tackling Energy Poverty Through Collective Advisory Assemblies and Electricity and Comfort Monitoring Campaigns. *Sustainability* 13 (17): 9671.

Partridge, T. 2012. Organizing Process, Organizing Life: Collective Responses to Precarity in Ecuador. *Interface: A Journal for and About Social Movements* 4 (2): 310–316.

———. 2016. Water Justice and Food Sovereignty in Cotopaxi, Ecuador. *Environmental Justice* 9 (2): 49–52.

———. 2017. Unconventional Action and Community Control: Rerouting Dependencies Despite the Hydrocarbon Economy. In *ExtrACTION: Impacts, Engagements and Alternative Futures*, ed. K. Jalbert, A. Willow, D. Casagrande, and S. Paladino, 198–210. New York: Routledge.

———. 2018. The Commons as Organizing Infrastructure: Indigenous Collaborations and Post-neoliberal Visions in Ecuador. In *The Right to Nature: Social Movements, Environmental Justice and Neoliberal Natures*, ed. E. Apostolopoulou and J. Cortes-Vazquez, 251–262. London: Routledge.

———. 2020. "Power farmers" in North India and New Energy Producers Around the World: Three Critical Fields for Multiscalar Research. *Energy Research & Social Science* 69 (101575).

———. forthcoming-a. The Right to Energy: Learning from Struggles for Food, Water, and Rights to Nature. In *Handbook on Energy Justice*, ed. S. Bouzarovski, S. Fuller, and T. Reames. Cheltenham: Edward Elgar.

———. forthcoming-b. Peripheral Labour: Work-Related Impacts and Inputs for Solar Generation in Uttarakhand, India. In *Capitalizing on the Sun: Critical Perspectives on the Global Solar Economy*, ed. J. Cross, D. Mulvaney, and B. Brown. Baltimore: Johns Hopkins University Press.

Partridge, T., J. Barandiaran, C. Walsh, K. Bakardzhieva, L. Bronstein, and M. Hernandez. 2020. California Oil: Bridging the Gaps Between Local Decision-Making and State-Level Climate Action. *The Extractive Industries and Society* 7 (4): 1354–1359.

Patel, R., and J.W. Moore. 2017. *A History of the World in Seven Cheap Things: A Guide to Capitalism, Nature, and the Future of the Planet.* University of California Press.

Pellow, D.N. 2018. *What Is Critical Environmental Justice?* Cambridge: Polity Press.

Pellow, D.N., and R.J. Brulle. 2005. Power, Justice, and the Environment: Toward Critical Environmental Justice Studies. In *Power, Justice, and the Environment: A Critical Appraisal of the Environmental Justice Movement*, ed. D.N. Pellow and R.J. Brulle, 1–19. The MIT Press.

Pollin, R., and Callaci, B. 2016. A Just Transition for U.S. Fossil Fuel Industry Workers. *The American Prospect*, July 6.

Powell, D.E. 2018. *Landscapes of Power: Politics of Energy in the Navajo Nation.* Durham: Duke University Press.

Rabinbach, A. 1992. *The Human Motor: Energy, Fatigue, and the Origins of Modernity.* Berkeley: University of California Press.

Ray, S.J., and J. Sibara, eds. 2017. *Disability Studies and the Environmental Humanities: Toward an Eco-Crip Theory.* Lincoln: University of Nebraska Pres.

Razack, S.H. 2002. *Race, Space, and the Law: Unmapping a White Settler Society.* Toronto: Between The Lines.

Reames, T.G. 2016. Targeting Energy Justice: Exploring Spatial, Racial/Ethnic and Socioeconomic Disparities in Urban Residential Heating Energy Efficiency. *Energy Policy* 97: 549–558.

Rees, W. 2019. End Game: The Economy as Eco-catastrophe and What Needs to Change William E. Rees. *Real-World Economics Review* 87: 132–148.

Rodríguez-Labajos, B., I. Yánez, P. Bond, L. Greyl, S. Munguti, G.U. Ojo, and W. Overbeek. 2019. Not So Natural an Alliance? Degrowth and Environmental Justice Movements in the Global South. *Ecological Economics* 157: 175–184.

Schlosberg, D. 2013. Theorising Environmental Justice: The Expanding Sphere of a Discourse. *Environmental Politics* 22 (1): 37–55.

Scoones, I., M. Leach, and P. Newell, eds. 2015. *The Politics of Green Transformations.* Abingdon: Earthscan.

Sekulova, F., G. Kallis, B. Rodríguez-Labajos, and F. Schneider. 2013. Degrowth: From Theory to Practice. *Journal of Cleaner Production* 38: 1–6.

SER. 2004. *The SER International Primer on Ecological Restoration*. www.ser.org & Tucson: Society for Ecological Restoration International Science & Policy Working Group.

Seyfang, G., J.J. Park, and A. Smith. 2013. A Thousand Flowers Blooming? An Examination of Community Energy in the UK. *Energy Policy* 61: 977–989.

Shrivastava, A. 2017. *Development and Prakritik Swaraj [Presentation]*.

Smil, V. 2004. World History and Energy. In *Encyclopedia of Energy*, 549–561. Elsevier.

Smith, L.T. 1999. *Decolonizing Methodologies: Research and Indigenous Peoples*. London: Zed Books.

Somers, M. 2008. *Genealogies of Citizenship: Markets, Statelessness, and the Right to Have Rights*. Cambridge University Press.

Stevis, D., E. Morena, and D. Krause. 2019. Introduction: The Genealogy and Contemporary Politics of Just Transitions. In *Just Transitions: Social Justice in the Shift Towards a Low-Carbon World*, ed. E. Morena, D. Krause, and D. Stevis, 1–31. London: Pluto Press.

Stirling, A. 2014. Transforming Power: Social Science and the Politics of Energy Choices. *Energy Research & Social Science* 1: 83–95.

Strauss, S., S. Rupp, and T. Love, eds. 2013. *Cultures of Energy: Power, Practices, Technologies*. Walnut Creek: Left Coast Press.

Sze, J. 2020. *Environmental Justice in a Moment of Danger*. Oakland: University of California Press.

Sze, J., and E. Yeampierre. 2018. Just Transition and Just Green Enough: Climate Justice, Economic Development and Community Resilience. In *Just Green Enough: Urban Development and Environmental Gentrification*, ed. W. Curran and T. Hamilton, 61–73. New York: Routledge.

Táíwò, O.O. 2022. *Reconsidering Reparations*. Oxford University Press.

TallBear, K. 2019. Badass Indigenous Women Caretake Relations: #Standingrock, #IdleNoMore, #BlackLivesMatter. In *Standing with Standing Rock: Voices from the #NoDAPL Movement*, ed. N. Estes and J. Dhillon, 13–18. University of Minnesota Press.

Temper, L., F. Demaria, A. Scheidel, D. Del Bene, and J. Martinez-Alier. 2018. The Global Environmental Justice Atlas (EJAtlas): Ecological Distribution Conflicts as Forces for Sustainability. *Sustainability Science* 13 (3): 573–584.

The Red Nation. 2021. *The Red Deal: Indigenous Action to Save Our Earth*. Brooklyn: Common Notions.

Thomas, S., L. Brischke, J. Thema, and M. Kopatz. 2015. *Energy Sufficiency Policy: An Evolution of Energy Efficiency Policy Or Radically New Approaches?* Stockholm, Sweden: ECEEE.

TNI. 2020. *Just Transition: How Environmental Justice Organisations and Trade Unions Are Coming Together for Social and Environmental Transformation.* Amsterdam: Transnational Institute.

Torres, H.G., and E.C. Marques. 2001. Reflexões sobre a hiperperiferia: Novas e velhas faces da pobreza no entorno municipal. *Revista Brasileira de Estudos Urbanos e Regionais* 4: 49–70.

Trauger, A. 2014. Toward a Political Geography of Food Sovereignty: Transforming Territory, Exchange and Power in the Liberal Sovereign State. *The Journal of Peasant Studies* 41 (6): 1131–1152.

Velicu, I., and M. Kaika. 2017. Undoing Environmental Justice: Re-imagining Equality in the Rosia Montana Anti-mining Movement. *Geoforum* 84: 305–315.

Vemuri, A., and D. Barney, eds. 2022. *Solarities: Seeking Energy Justice.* Minneapolis: University of Minnesota Press.

Verhoeven, H. 2021. The Grand Ethiopian Renaissance Dam: Africa's Water Tower, Environmental Justice & Infrastructural Power. *Daedalus* 150 (4): 159–180.

Vía Campesina. 1996. *Food Sovereignty: A Future Without Hunger / The Right to Produce and Access to Land.* Presented at the World Food Summit (Rome), 11–17 November.

Voyles, T.B. 2015. *Wastelanding: Legacies of Uranium Mining in Navajo Country.* Minneapolis: University of Minnesota Press.

Walia, H. 2012. Moving Beyond a Politics of Solidarity Toward a Practice of Decolonization. In *Organize! Building from the Local for Global Justice,* ed. E. Shragge, J. Hanley, and A.A. Choudry. Oakland: PM Press. epub 345–363.

Watts, M., and R. Peet. 2005. Liberating Political Ecology. In *Liberation Ecologies: Environment, Development, Social Movements,* ed. R. Peet and M. Watts. London: Routledge.

Whyte, K.P. 2018. What Do Indigenous Knowledges Do for Indigenous Peoples? In *Traditional Ecological Knowledge: Learning from Indigenous Practices for Environmental Sustainability,* ed. M.K. Nelson and D. Shilling, 57–81. Cambridge University Press.

———. 2020. Too Late for Indigenous Climate Justice: Ecological and Relational Tipping Points. *Wiley Interdisciplinary Reviews: Climate Change* 11 (1).

———. 2021. Against Crisis Epistemology. In *Routledge Handbook of Critical Indigenous Studies,* ed. B. Hokowhitu, A. Moreton-Robinson, L. Tuhiwai-Smith, C. Andersen, and S. Larkin, 52–64. New York: Routledge.

Whyte, K.P., and C. Cuomo. 2017. Ethics of Caring in Environmental Ethics: Indigenous and Feminist Philosophies. In *The Oxford Handbook of Environmental Ethics,* ed. S. Gardiner and A. Thompson, 234–247. Oxford University Press.

Wildcat, D. 2009. *Red Alert! Saving the Planet with Indigenous Knowledge.* Colorado: Fulcrum Publishing.

Wilgosh, B., A.H. Sorman, and I. Barcena. 2022. When Two Movements Collide: Learning from Labour and Environmental Struggles for Future Just Transitions. *Futures* 137: 102903.

Winter, C.J. 2021. *Subjects of Intergenerational Justice: Indigenous Philosophy, the Environment and Relationships.* London: Routledge.

XSE. 2018. *Tenemos energía. Retos de la transición hacia la soberanía energética.* Barcelona: Icaria Editorial.

Yazzie, M.K. 2018. Decolonizing Development in Diné Bikeyah: Resource Extraction, Anti-Capitalism, and Relational Futures. *Environment and Society* 9 (1): 25–39.

York, R., and S.E. Bell. 2019. Energy Transitions Or Additions? *Energy Research & Social Science* 51: 40–43.

Zografos, C., and P. Robbins. 2020. Green Sacrifice Zones, Or Why a Green New Deal Cannot Ignore the Cost Shifts of Just Transitions. *One Earth* 3 (5): 543–546.

CHAPTER 3

A Critical Energy Research Agenda

Abstract This section maps out a Critical Energy Research agenda that analyzes energy relations in light of their role within broader systems of (re)production, (geo)politics, economics, and ecological regeneration. This requires moving beyond the popular 'energy justice' framework, both analytically and politically, and also continuing to engage with emergent branches of EJ such as Critical Environmental Justice studies and Multispecies Justice. Reorienting our work in these ways carries theoretical, methodological, and political implications. Accordingly, critical energy research also reconsiders the roles and responsibilities of those who write about injustice. Five key approaches constitute this agenda for critical energy research: (1) foregrounding the roles of EJ communities, Indigenous groups, and diverse social movements; (2) maintaining a critical focus on systems of production and extractivism (and on how ongoing processes of settler colonialism and multiple forms of exploitation interact and underpin energy-intensive systems of capitalist production); (3) reframing transitions as collaborative projects of justice; (4) drawing on restorative environmental justice to see justice as the restoration of mutually supportive relationships between beings of all kinds; (5) renewing practices of reflexivity by subjecting our own analyses, commitments, research relations, and notions of solidarity to ongoing political scrutiny.

Keywords Critical environmental justice • Multispecies justice • Reflexivity • Solidarity • Refusal • Critical energy research

© The Author(s), under exclusive license to Springer Nature Switzerland AG 2022
T. Partridge, *Energy and Environmental Justice*,
https://doi.org/10.1007/978-3-031-09760-7_3

Introduction: Beyond the 'Energy Justice' Framework

Energy-related injustices have always been an area of focus for the EJ movement, both globally and within the US. The history of environmental justice—as a concept, mobilizing force, and a response to environmental racism—is traced back through multiple sites of environmental conflict and struggles for social justice. A prominent case relates directly to the operation of energy systems. In 1978 in North Carolina, in the town of Afton—a predominantly African American town in Warren County, one of the poorest counties in the state—residents lay their bodies down on roads leading to the local landfill to stop trucks from dumping more than 32,000 cubic yards of soil contaminated with highly toxic PCBs (polychlorinated biphenyls) (Bullard 1990, p. 30). Used as a heat transfer fluid in electrical transformers, the PCBs originated as toxic waste from the operations of Ward Transfer Company (ibid.). Over six weeks, more than 500 people were arrested during local actions and nonviolent street protests (Skelton and Miller 2016). Despite their resistance, the waste was eventually dumped in that landfill site. But these actions garnered national attention, generated support for parallel efforts elsewhere, and led researchers to identify and document overlapping national patterns of environmental racism—as detailed in the 1987 Commission for Racial Justice report, "Toxic Wastes and Race in the United States," and in countless projects since (ibid.). Energy research, however, has hardly kept up.

While Native scholars and activists and their collaborators have for a long time addressed the justice-implications of different energy systems and energy relations, tracking the systematic and violent targeting of Indigenous peoples (e.g., in North America: LaDuke 1981; Gedicks 1993; Smith and Frehner 2010), the 'energy research' literature has by and large followed a different track—mostly considering only questions of technology and finance rather than justice. Anthropologist Laura Nader describes her career of more than four decades engaging with energy professionals and decision-makers as a series of frustrations: the majority of her interlocutors still to be convinced that energy policy is more than a technological problem (Nader 2010). These gaps in policy spheres and throughout the literature are there because, for decades, energy research did not engage closely with EJ as a field, nor with diverse EJ communities and Indigenous groups who were at work identifying, exposing, challenging, and fighting against multiple energy-related injustices. The work of movements was largely disregarded. These oversights are particularly indefensible given how hard EJ movements and communities have fought to have their stories heard around cases of energy injustice.

Before being executed by the Nigerian military regime in 1995 along with eight other members of the Movement for the Survival of the Ogoni People (MOSOP)—a governmental attempt to silence popular resistance to decades of crimes—Ken Saro-Wiwa documented the environmental devastation caused by Shell's oil extraction operations in the Niger Delta and their corporate complicity in state-sanctioned killings and torture (Saro-Wiwa 1995). Amnesty International has compiled extensive evidence of these crimes and has called for a criminal investigation (AI 2017). The resistance movement continues today, connecting and sharing information and strategies with other EJ communities around the world (Temper 2013; Ngwakwe 2021). In parallel efforts in Ecuador, Indigenous activists and allies have long been fighting to hold oil companies to account for the suffering they have inflicted upon Indigenous communities (Sawyer 2004). Their struggle, too, has been addressed by a growing network of international groups. A recent *Stand.earth* report found that found 89% of all the oil drilled and exported from Amazon regions comes from Ecuador; half of that total goes to California; and half of that amount goes to just three refineries in the Greater Los Angeles area, undermining the often-touted self-image of that state as a climate leader (Robertson 2021). Many other global interconnections and implications of energy systems are being documented in increasing detail by EJ actors and mobilizations. Energy research still has much to learn by engaging in detail with these diverse justice struggles.

Chapters 1 and 2 of this book show how placing critical EJ insights and approaches at the heart of multidisciplinary energy research can deepen those processes of learning, interaction, and solidarity that support renewed connections between multiple movements and researchers. Chapters 1 and 2 also highlight the range and depth of critical thinking that energy research can draw on in order to sharpen the analytical tools we use to better understand energy relations and their implications past, present, and future. Refocusing energy research in these ways has aspects that are strategic or practical or which, at least, concern the kinds of ideas and initiatives that energy research will explore in support of transitions recast as collaborative projects of justice: committing to pluralism, diversity, degrowth, and restorative environmental justice. Connecting all these positions are the theoretical insights and orientations that cast *energy* itself in a new light: reassessing what we mean by 'energy' when we seek to more effectively scrutinize its roles within the power structures, social relations, and institutional configurations that produce environmental and social injustices (Harvey 1996). This is the work of politicizing energy histories; interpreting the typically genericized 'energy' as referring to the potential and practice of commodifiable work; centering critiques of racial capitalism and settler colonialism;

and analyzing energy relations through their roles within systems of production, extraction, and exploitation (or, conversely, within efforts to restore socio-ecological relations of respect and reciprocity).

Building on these perspectives and approaches—and their theoretical, political, and methodological implications—the rest of Chap. 3 maps out an agenda for critical energy research. First, this requires that we move beyond the popular 'energy justice' framework, both analytically and politically. Second, this requires connecting energy research with critical perspectives, insights, and approaches from within emergent branches of EJ, including Critical Environmental Justice studies and Multispecies Justice. A further component of critical energy research concerns how we think about research relations and practices—specifically by making explicit the fact that how we think about justice has implications for how we conceive, conduct, and write about research processes. If we consider justice not as a set of ideal principles that researchers are hopeful will be applied by governments and corporations once they have had the error of their ways pointed out to them by incisive scholarship, but instead consider justice as emergent within the claims and praxis of groups who are already fighting against the different injustices that shape their everyday experiences—this is an orientation that rightfully demands of researchers different forms of engagement and reflexivity. A first step here involves looking again at different approaches to justice within 'energy justice.'

Initial framings of energy justice in the academic literature focused largely on differential access to the benefits of energy systems (Bickerstaff et al. 2013; Day 2020). Now the term is applied more widely, but most energy justice scholarship still focuses on policy assessment and bottom-line economics: assuming an anthropocentric perspective and seeking to address injustice by petitioning state and corporate actors to 'do the right thing' in terms of policy design, ethical orientation, and financial investment (e.g., Sovacool et al. 2016). Pursuing these goals, energy justice has adhered to a framework of three 'tenets': distributional, procedural, and recognition justice (Hess and Ribeiro 2016). This framework, first developed within EJ, addresses the uneven distribution of environmental harm, risk, and protection across societies as well as disparities in which voices are recognized and respected within decision-making processes (Bullard 1996; Schlosberg 2007). EJ work (and latterly energy justice work too) has since expanded upon these principles by drawing on theorists including Sen and Nussbaum to examine how environmental injustices impinge upon the capabilities, capacities, and flourishing of different communities (ibid.). Beyond these theoretical approaches, however, the energy justice literature largely obscures or misunderstands its relation with EJ movements, anti-racism organizing, histories of direct action, and ongoing Indigenous struggles. Insights are routinely overlooked concerning racialized production, settler colonialism, and the many direct links between increasing global energy demand and exacerbated socio-ecological injustices. Critical approaches to energy research, by contrast, seek to foreground EJ contributions and to reconnect 'energy justice' with its Indigenous, anti-racist, activist roots.

An early example of the energy justice term being mobilized is the 1999 foundation and ongoing work of the Energy Justice Network in the US. Their mission statement emphasizes support for communities negatively affected by energy systems and draws directly on the Environmental Justice Principles adopted at The First National People of Color Environmental Leadership Summit (POCELS 1991). The Network advocates for deep reductions in energy consumption, a rapid phase-out of nuclear power and fossil fuels, and in-depth critiques of proposed 'solutions' to contemporary energy dilemmas (Energy Justice Network n.d.). Hernández (2015) has since incorporated many of these principles (and their activist orientation) into a rights-based model for energy justice. The (1991) EJ Principles affirm a number of demands that go beyond standard definitions of distributional, procedural and recognition justice but which are often overlooked in energy research. These include strict accountability for [energy] producers (#6) and the right of those suffering environmental injustice to full compensation, reparations and health care (#9); a focus on employment conditions and workers' rights (#8); an approach to procedural justice that extends beyond 'meaningful participation' to demand rights as "equal partners at every level of decision-making" from planning and implementation through to, crucially, enforcement and evaluation (#7); and the view that energy systems are not fully addressed only (nor even primarily) as matters of policy but are instead deeply social processes (#17) which further require societal education about diversity and experiences of inequality (#16).

The most visible energy justice scholarship pays little attention to these activist histories and continues to narrate a history of the concept that focuses primarily on academic literature after 2013 (e.g., Jenkins et al. 2021). What have become canonical works routinely make no reference to prior and ongoing Indigenous struggles for energy justice—even those that have mobilized specifically around (and published on) 'energy justice' years before the term gained traction in academia (e.g., EJNA 2009; Goldtooth and Awanyanka 2010). See Images 3.1 and 3.2 for a poster by

TAKE ACTION!

Now is the time. Tribally owned and operated renewable energy along with green jobs that help reduce dependence on fossil fuels are central to a sustainable and affordable low-carbon energy future. Energy decisions made in the coming months by Congress and the Obama administration will have significant implications for many generations to come.

BE A VOICE FOR THE EARTH

INDIVIDUALS — Contact your Senators and Member of Congress and urge them to: "Support renewable energy and efficiency in Native communities. We want Green Jobs for Native Peoples Now!" **CALL:** (202) 224-3121. It's best to make three quick calls, one for each of your two Senators and one for your Member of Congress. Please call now!

ORGANIZATIONS AND TRIBAL GOVERNMENTS —The coalition of Honor the Earth, Intertribal Council On Utility Policy, Indigenous Environmental Network, and the International Indian Treaty Council collaborated to create the Energy Justice in Native America Policy Paper for the Obama Administration and the Congress. The paper has a detailed strategy for working toward energy justice and green economies in Native communities. The paper has a detailed strategy for working towards energy justice and green economies in Native communities; the full version is available at **honorearth.org**. Please stand with us and sign onto the paper for upcoming submission to key decision-makers! Email **greeneconomy@honorearth.org** today to add your organization or tribal government to our growing list of allies.

♻ 100% PCW recycled   "Warrior Wombyn" Art by: Votan Ik'ahn(Maya-nahua) www.insurgentes.net

GREEN ECONOMIES IN NATIVE COMMUNITIES: MASSIVE POTENTIAL, MAXIMUM IMPACT

A green jobs economy and a new, forward thinking energy and climate policy will transform tribal and other rural economies, and provide the basis for economic recovery in the United States.

For every dollar invested, renewable energy development creates more jobs than fossil fuels like coal, oil, or gas.

Renewable energy is energy security: Unlike the volatile prices of fossil fuels, the cost of wind and solar resources can be projected into the future, providing a unique opportunity for stabilizing an energy intensive economy.

ADDRESSING CLIMATE CHANGE AND BOOSTING ECONOMIES: RENEWABLE ENERGY AND EFFICIENCY IN NATIVE AMERICA

To address the climate change crisis, the US will need to install thousands of megawatts of new clean energy generation over the next decade, and we will need to get efficient.

One scenario is developing 125,000 megawatts of wind power in the next 10 years (an estimate of the amount needed to stabilize the country's carbon emissions). This scenario would create about 400,000 new manufacturing jobs!

Efficiency also creates needed jobs. Conservation in housing and other buildings could reduce our power use by up to one-third and reduce the level of fuel poverty in our communities.

Tribal communities have tremendous untapped wind and solar energy resources and are central to the green economy of the future. Our tribal lands have:

- 535 Billion kWh/year of wind power generation potential, about 14% of US annual generation and;
- 17,000 Billion kWh/year of solar electricity generation potential, about 4.5 times total US annual generation.

Image 3.1   2009 poster by Energy Justice in Native America (Part one)—an example of the 'energy justice' concept being developed and deployed in Indigenous activism long before it became the focus of an academic (sub)field. "Warrior Wombyn (aka Rezzie The Riveter)" artwork © Votan Ik NSRGNTS. COM. Materials used with permission

A GREEN ECONOMY FOR THE SEVENTH GENERATION

NOW IS THE TIME TO ACT!

ECONOMIC STIMULUS BILLS. ENERGY LEGISLATION. CLIMATE CHANGE INITIATIVES.

All of these will move through the Congress quickly. Native peoples demand a place at the table.

Historic and present impacts of climate change and energy development on American Indian reservations and Alaska Native villages are devastating.

Today, we have the potential to catalyze a green economy, with justice for our Native communities.

HONOR THE EARTH

INTERTRIBAL COUNCIL ON UTILITY POLICY

INDIGENOUS ENVIRONMENTAL NETWORK

INTERNATIONAL INDIAN TREATY COUNCIL

TO BEGIN WORKING TOWARD ENERGY JUSTICE WE URGE THIS ADMINISTRATION TO:

- Increase the capacity of tribal education institutions to train the next generation of green job workers and continue to boost the capacity of technical training programs.
- Create financial support for efficiency in federal fuel assistance programs, and for the installation of low cost energy savers like solar heating panels.
- Ensure tax credits and financing for renewable projects are applicable to tribally-owned renewable energy development.
- Ensure priority access to the electrical grid for green energy.
- Ensure international debt reduction programs, to reduce the pressure on governments and Indigenous communities to cut down their forests or sell their fossil fuels. Support preservation of biodiversity and Indigenous rights in these areas.
- Create a moratorium on off shore oil leasing in Alaska.
- Not fund or provide incentives for "clean" coal or nuclear power. Neither are solutions to climate change and neither represent sustainable economic development for Native communities.

> One example of green economic development is to fully authorize a tribal solar project to cover the 355 miles of an open federal Central Arizona Project canal with solar photovoltaic cap to generate over 1500 megawatts of clean, efficient solar power displacing dirty coal.

WITH THESE ACTIONS, WE CAN DEVELOP A GREEN ECONOMY BASED ON JUST RELATIONS WITH NATIVE AMERICA.

Image 3.2 2009 poster by Energy Justice in Native America (Part two)—an example of the 'energy justice' concept being developed and deployed in Indigenous activism long before it became the focus of an academic (sub)field. "Warrior Wombyn (aka Rezzie The Riveter)" artwork © Votan Ik NSRGNTS. COM. Materials used with permission

Energy Justice in Native America that accompanied the release of their collaborative 2009 report on energy production and efficiency, resource extraction, Indigenous rights, and environmental justice.

The omission of Indigenous struggles for energy justice within much of the scholarly literature further marginalizes already subjugated lives and knowledge, discounting valuable insights from first-hand experiences for identifying and counteracting energy-related injustices (Montoya 2016; Estes 2019). As a result, efforts to address and prevent environmental harms are minimized in their scope. Analytically, the omission reflects a problematic assumption: that measures for procedural justice can be implemented within contemporary political and judicial systems, including those that remain rooted in colonial relations. That assumption overlooks Indigenous critiques of such processes (see Chap. 1: Settler Colonialism and Sovereignty), particularly those concerning the shortcomings of participation and recognition as 'tenets' of justice (Goldtooth 1995; Ranco 2008; Whyte 2017). Consequently, much of the energy literature further entrenches an institutional perspective on social change, downplaying (and misunderstanding) the role of social movements.

As discussed above (Chap. 1: Autonomy, Prefiguration, and the State), justice movements often initiate broader societal change by enacting in the present the kinds of socio-environmental relations that are needed in order to move beyond current political-economic constraints and to halt dominant practices of environmental destruction. Rooting the energy justice concept in the contexts from which it emerged—in the ideas and praxis of Indigenous and activist groups—means defining justice in light of the demands for which diverse movements are already fighting. While almost a decade of academic literature has sought to refine energy justice and its potential application within current (and highly unequal) structures of ownership, control, and decision-making, Indigenous and other resistance groups have fought for many years—in many cases, for centuries—to overcome those structures and to defend diverse lives, ecologies, and ways-of-being. Justice objectives are served effectively when specific campaigns and acts of resistance are supported. Powerful actors responsible for the (re)production of energy and environmental injustices tend to change their behavior not when they are asked to do so but when they are made to do so.

Critical energy research draws on the work of scholars who are already challenging the energy justice framework: e.g. by analyzing how multiple types of injustice interconnect (Finley-Brook et al. 2018); re-articulating justice in terms of care relationships (Damgaard et al. 2022); reassessing the framework's political assumptions (Pellegrini-Masini et al. 2020); emphasizing sensitivity to political context (Hernández et al. 2022); critiquing the framing of injustice primarily as an abstract problem (Jasanoff 2018; Szolucha 2018); or highlighting how injustices are (re)produced through the normalization of pollution, the commodification of nature, technocratic commitments to energy centralization, and reliance on top-down decision-making processes (Lee and Byrne 2019). Others note that much of the framework-based literature fails to engage with either (environmental) racism and privilege (Ciplet 2021; Newell 2021) or the history of justice within struggles for Civil Rights (Galvin 2020). Further gaps in the literature reflect a lack of engagement with postcolonial critiques of energy and development (Castán Broto et al. 2018); notions of justice emergent within the socio-ecological commons (Yaka 2019); and transformative, decolonial approaches to EJ implementation and epistemic justice (Temper 2019; Avila et al. 2021). No single project can simultaneously engage, in any detail, with these many different approaches. But what critical energy research *can* do is first highlight (rather than obscure) relevant and complimentary perspectives, then work to build solidarities with those whose research and on-the-ground actions have helped generate these ideas. The goal remains to continue developing a closer (and more mutually supportive, exploratory, and interconnected) set of relationships between energy research and EJ.

Critical energy research thus strongly rejects the claims made by Jenkins (2018) and others that energy justice has a 'non-activist' origin and that avoiding association with apparently 'anti-establishment' actors is a strategy to be embraced in order to more easily contribute to 'mainstream' policy-making. Critical approaches instead recognize that pushing for meaningful societal change cannot be limited to action solely within extant political, judicial, and analytical frameworks: "the fight against injustice is not necessarily the same as outlining some positive conception of justice" (Healy and Barry 2017, p. 452). The 'justice' in energy justice is either informed by ongoing collective struggles or else it fails to address systemic injustices and ultimately serves the interests of governments and energy corporations by promoting reformist measures rather than structural change. Learning from EJ communities, Indigenous struggles, and engaged scholarly approaches, critical energy research moves beyond the energy justice framework by re-framing energy justice as a spectrum of political engagements within broader struggles for transformative social change.

CRITICAL ENVIRONMENTAL JUSTICE

The analyses of themes addressed in Chaps 1 and 2—diversity and plurality, direct democracy, racial capitalism, settler colonialism, the state and autonomy—are informed by Critical Environmental Justice (CEJ) studies. CEJ illustrates in detail how these issues interconnect and how critical analysis can support diverse actions for radical, transformative change. The growth of EJ as a field has been traced from an initial focus on distributive justice, to later adding participation, representation, recognition, and capabilities as key concerns, and then also drawing on justice theorists such as Young (1990) and Fraser (2000) to examine why judicial systems routinely fail to deliver justice and why injustices emerge in the first place (Schlosberg 2007). EJ has also expanded in scope geographically (outside the US and globally), thematically (e.g. chains of production, life-cycle analyses), and in terms of its subjects (e.g. non-humans), as well as through a focus on community-led science and participatory action across the global environmental justice movement (Walker 2012; Schlosberg 2013; Martinez-Alier et al. 2016). Acknowledging these developments, CEJ itself builds on and further expands EJ research by reconceptualizing how inequalities and different forms of racist violence are produced and resisted (Pellow and Brulle 2005; Holifield et al. 2010; Adamson 2011). CEJ perspectives offer vital guidance for critical energy research.

As described elsewhere (Partridge 2020), CEJ identifies approaches that are particularly applicable to the study of energy transitions and the multiscalar implications of novel, distributed, and disruptive models of energy generation. David Pellow (2018) describes four 'pillars' of critical environmental justice: (i) attention to how multiple categories of social difference and marginalized positions *intersect* in the production and experience of injustice (categories including race, gender, sexuality, ability, class, and species, both human and beyond-human); (ii) *multiscalar* methodological and theoretical approaches to the complex 'causes, consequences, and possible resolutions' of injustices across both spatial and temporal scales; (iii) an anti-authoritarian orientation that recognizes social inequality as reinforced by state power and pushes for *transformative* (more than reformist) responses to injustice; (iv) a focus on the *indispensability* of all humans and more-than-human actors and their necessary roles in building our shared futures (working against practices that render certain populations expendable) (Pellow 2018, pp. 17–18, emphasis added). The four pillars of CEJ both reflect and inform core aspects of restorative environmental justice, as described in Chap. 2: (i) pursuing solidarity and radical inclusivity across multiple global movements and contexts (multiscalar and transformative work); while also (ii) articulating how diverse, complex identities both emerge from and are shaped by overlapping socio-environmental injustices: identities and experiences which are all, individually and together, vital in confronting those injustices (intersectionality and indispensability). Each of these aspects, or pillars, contributes to collective efforts to restore relationships of mutual care and reconnection between humans and ecosystems.

Crucially, putting indispensability into practice is not just a call to restore less harmful and more respectful socioecological relationships across species, it also forms a deliberate organizing strategy: "In 2002, at the Second People of Color Environmental Leadership Summit, activists penned and endorsed a document: The People of Color Environmental Justice 'Principles of Working Together.' One of those principles [3.A] reads, 'The Principles of Working Together recognize that we need each other and we are stronger with each other. This Principle requires participation at every level without barriers and that the power of the movement is shared at every level'" (POCELS 2002, at Pellow 2018, p. 40). Indispensability therefore informs how energy (and other) transitions are devised, designed, and enacted. Building transitions as collaborative projects of justice for the future—and as opportunities for transformative social change—requires (i) scrutiny of the systems of power that have had an outsize influence on shaping the contemporary world (including logics of capitalism, unequal overconsumption, and economic growth), coupled with (ii) active engagement across multiple scales to identify and support alternative strategies already underway and in formation (Ryder et al. 2021, p. 156). These alternative strategies include movements for energy sovereignty and energy localization initiatives.

While all four CEJ pillars are interconnected—the emphasis in those connections shifting across contexts and within different moments of struggle—the concept of indispensability and the demand for transformative responses to injustice are closely related within transitions design and praxis. In the case of energy localization projects, this might mean specifically incorporating more inclusive, participatory, and autonomous structures of governance. We see this when the design and management of energy systems are subject to context-specific forms of direct democracy and accountability so as to better provide equitable access to energy and resources (Partridge 2020). Such initiatives seek to build, and to build on, modes of organizing that strategically minimize their dependence on the relationships of inequality and exclusion commonly associated with state institutions. Within resistance praxis, indispensability and transformative action can also connect when groups use citizen science techniques (such as counter-mapping) to challenge institutional accounts and to reverse the spatial erasure of human and nonhuman kin; these strategies are described in ann-elise lewallen's engaged work with grassroots, land-defense, and anti-nuclear groups in Japan and India (v. NNAF 2019).

Maintaining a focus on how histories of domination and exclusion have created (and continue to exacerbate) environmental injustices, CEJ approaches articulate the need for structural social and political change. This orientation connects multiple social movements and raises a number of questions that can be used to orientate energy-focused research (Partridge 2020): How do other diverse struggles over land-use, labor, and livelihoods influence contestation surrounding new energy projects? How are energy-related injustices produced within (or are part of upholding) unequal political and economic structures? What sociopolitical histories have marginalized certain groups and how have diverse social movements mobilized to counteract those forces?

Another crucial insight from CEJ for energy research is to adopt an expansive view of environmental justice: to consider how the CEJ framework might productively be applied in a wider range of cases wherever "humans and nonhuman natures are bound up in the struggle for environmental justice against myriad threats at multiple scales" (Pellow 2018, p. 44). For energy research, this runs counter to a tendency in the literature to try and separate issues of energy justice from the fields of environmental justice and climate justice. Such attempts try to isolate energy-related injustices from broader struggles for overlapping forms of justice. This might be done as a legitimate way of focusing research questions and practices, adhering to the caution that EJ risks losing its explanatory and mobilizing power if it is used to analyze or explain "every problem at the intersection of development and social inequality" (Pellow and Brulle 2005, p. 16). Drawing on CEJ, however, critical energy research can pursue a focus on the *specificities* of energy systems and energy relations—the diverse materialities, processes, and socioecological relations involved—as a way of also reassessing and reconsidering the broader social, political, and environmental contexts within energy systems operate. Focusing on these complex connections, critical energy research offers another route toward engaging with (and contributing to) parallel justice-oriented struggles and investigations.

CLIMATE, WATER, CULTURAL JUSTICE

In addition to the emergence of Critical Environmental Justice studies, the evolution of EJ is reflected in what Walker (2012) identifies as horizontal and vertical diffusions: involving new places and contexts as well as new scales, concerns, and movements. The concept of climate justice, for example, actively pursued by a global movement of movements with diverse histories and experiences, highlights the planetary scale of environmental injustices in a rapidly warming (and climate-disrupted) world. Achieving climate justice would mean limiting exposure to the very unevenly experienced socio-ecological disruptions caused by climate change as a result of atmospheric pollution while also recognizing the very unevenly distributed responsibilities for creating that pollution (Newell and Mulvaney 2013). More than 20 years ago, the need for radical changes to energy and economic systems was already being framed in terms of climate justice. Even so, the powerful corporations identified as leading culprits driving global warming have found ways to not only continue operating but also to continue growing since then. This is despite (but also, as companies seek to extend their final days as long as possible, because of) something that has been clear for decades: "the severity and planet-wide nature of climate change represents a sort of an endgame for the global oil corporations" (Bruno et al. 1999, p. 5). Climate justice has gained ground as a mobilizing platform through the actions of multiple networks, initially including 'Climate Justice Now!' (at the COP-13 negotiations in Bali in 2007) and 'Climate Justice Action' (around the COP-15 negotiations in Copenhagen in 2009):

> "For many… the broad position underlying the use of the term [climate justice] is the politicisation of climate change—understanding that it results from our current and historical social relations, and that in order to address it we need fundamental changes to our economic and political systems. The 'injustice' is that the industrialised western world are disproportionately (both historically and currently) responsible for the emissions that are causing climate change, and are now using it as an excuse to accumulate further through the implementation of market-based false solutions. Meanwhile, the geographical and political south will suffer the worst effects of climate change; their territories and resources were plundered and polluted to feed western industrialisation, and now climate change is being used as an excuse to colonise, privatise and dispossess further through the creation of new markets to 'solve' the crisis". (Building Bridges Collective 2010, p. 27)

Another example of a connected, but distinct, branch of environmentally-focused, justice-oriented activism and literature relevant to energy research falls under the term 'water justice.' Frequently undertaken through global collaborations, water justice research engages justice theory and praxis through social struggles and emancipatory action (e.g., Vila Benites and Bonelli 2017). Considered as one dimension of a larger vision for EJ, water justice addresses the specificities of water allocation and management through a deeper understanding of the materiality and politics of water (e.g., Bakker 2012). The evolution of both climate and water justice debates has brought a number of ideas and methods into the broader EJ field, including critical perspectives on knowledge production; development and post-development; infrastructure ownership and control; and strategies for democratizing resource governance (Perreault et al. 2018). Of particular importance for the project of articulating a critical energy research agenda are perspectives on relationality and cultural justice.

Theories of justice typically present normative descriptions of ideal principles rather than studying specific experiences of (in)justice and effects on wealth and authority distribution (Joy et al. 2014, p. 965). Alternative approaches scrutinize (in)justice in relation to conflicting notions of rights and control, studying how inequalities come to matter differentially through social relationships and practices (Walker 2012). These alternative approaches, widely applied in water justice work, conceptualize justice as "relational, situated, and context-sensitive" (Roth et al. 2014, p. 949). While situated—that is: defined by, and responsive to, particular understandings of justice within certain sociocultural, political, and geographical contexts—such understandings of (in)justice are not limited to isolated cases. Rather than considering place-based struggles as bounded cases of political action (Harvey 1996), relational approaches account for the continuous relationships of interaction and co-constitution that link place, subaltern spaces of politics, and global or universal political imaginaries (Massey 1999; Featherstone 2005, p. 252). This holds relationships between particularity and plurality in productive tension—a tension used strategically by environmental justice movements and documented by EJ scholars (Martinez-Alier et al. 2014, 2016). For example, a relational understanding of multi-scalar, inter-species, and intergenerational connectedness enables recognition of plural justice norms across diverse contexts; at the same time, broader (or global) concepts of justice can equip different groups with "an important vocabulary" in specific situated (or local) struggles against inequalities and dispossession (Temper and Del Bene 2016, p. 41). Injustices come to light—and come to mat-ter—through these processes of negotiation and contestation.

In order to fully engage with diverse, situated experiences of oppres-sion, relational approaches to justice require a further analytical step be taken: challenging models of justice that assume equality among diverse actors. Systemic exclusion and domination are often based on the rhetoric of universality, with the result that minority views are obscured: "homoge-neous concepts of justice based on abstract, universal criteria tend to poorly correspond (and respond) to the experiences of and claims made by the 'non-equals': marginalized indigenous and peasant societies, for instance, or women" (Zwarteveen and Boelens 2014, p. 147). Accounting for this analytically requires extending 'recognition' justice to also address questions of authority, legitimacy, and cultural-political organization—what has been referred to as "cultural justice" (Boelens and Seemann 2014).

Cultural justice disrupts the uniformity favored by corporations and governments that denies (even if it cannot erase) locally and historically specific organizational forms—this means recognizing legal pluralism (where diverse forms of ancestral, judicial, market, development, and inter/national law all differentially co-exist) and exposing the fallacy of the idea of an equitable legal 'playing field' shared among corporations, com-munities, and marginalized individuals (Boelens and Seemann 2014; Perreault et al. 2018). Critical EJ scholars describe these actions as dis-rupting and decolonizing the idea of 'participatory parity' (Young 2000; Temper 2019). As Shuar, Kichwa, and other Indigenous groups have experienced in Ecuador defending their lands in national courts against oil companies routinely circumventing regulations, the legal arena is regularly structured and/or manipulated such that human rights are subordinate to the interests of economic and political elites (Figueroa 2006). Cultural justice involves not only upholding minority rights but, furthermore, actively challenging institutions and powerful actors who threaten the diverse ways-of-living and ways-of-relating that more comprehensive con-stitutional rights are, at least nominally, designed to protect.

Cultural justice also corresponds with calls to expand the concept of recognition by distinguishing between dominant, institutional definitions and practices of inclusion from those that are generated by groups actively involved in justice-oriented mobilizations (Pallares 2002; Partridge 2015). For example, studies of "cultural-political projects" show cultural justice for marginalized groups can involve both (i) refusing to accept categories imposed upon them and conditions that reflect "prevailing power relations," and (ii) pursuing ways to "react to, modify and strategically use the ruling symbolic order" (Hidalgo et al. 2017, p. 72). Cultural justice thus examines not only practices and challenges of recognition and marginalization but also the "right to be different" or the "right to have rights" (Perreault et al. 2018, pp. 36–37). Such a formulation of cultural justice demands differences in priorities and practices be not only respected but also, as may be required, actively supported and protected from processes that would otherwise impinge upon them. This aligns with a core message of the founding Principles of Environmental Justice: a need to secure the "political, economic and cultural liberation that has been denied for over 500 years of colonization and oppression" (POCELS 1991). In Chap. 2, above, pluralism and diversity were shown to be integral to the collaborative work of building transitions as transformative projects of justice. By taking seriously the issues of difference and relationality that cultural justice brings to the fore, critical energy research can more effectively contribute to those efforts.

Multispecies Justice

In addition to those framed in terms of cultural justice, EJ perspectives that problematize, expand, or reject the concept of recognition include perspectives centered around value pluralism. In these cases, the defense of resources and ecosystems is not reduced exclusively to economic valuation but instead acknowledges diverse livelihoods, forms of sacredness, aesthetic values, and cultural values—while always challenging those who claim the power to impose a singular 'bottom line' language of valuation upon those plural values (Martinez-Alier 2002; Zografos and Martínez-Alier 2009). Extending this resistance to singular valuation languages, there are groups who take further steps in order to pursue justice through resistance to singular notions of existence, acknowledging instead multiple worlds or ontologies.

Ontologies are understood as "total enactments involving discursive and non-discursive aspects… about what kinds of things do or can exist, and what might be their conditions of existence" while conflicts between differently-enacted worlds emerge as they interact, intermingle, and "strive to sustain their own existence" (Blaser 2009, p. 877). Some environmental conflicts are therefore more accurately seen as ontological conflicts that involve distinct configurations of relations between humans and nonhumans; recognizing and respecting these relations means not imposing a singular understanding of what exists and how (Cadena and Blaser 2018). For example, when actions taken against mining projects in defense of particular mountains or lakes is not reduced to being an expression of Indigenous "beliefs" about "sacred beings" but rather recognizes that some worlds live "partly outside the separation between nature and humanity" (Escobar 2020, p. xiii). At stake in such conflicts is the very existence of those multiple worlds.

Acknowledging, and engaging with, the basis of ontological conflicts—the presence of multiple worlds within worlds—requires a reassessment of what counts as 'justice' within environmental justice and of who counts among those who might suffer less through the realization of justice objectives. While much EJ work to date has largely focused on environmental harms endured by people whose human rights and civil rights are disregarded by powerful and industrial actors, a growing body of research examines environmental injustices that affect beings of many different kinds: beings whose rights to an unhindered existence, free from suffering, are yet to be fully recognized or understood (Gruen 2014; Pellow 2014; Gaard 2017). To this end, critical EJ approaches explicitly seek to understand inequalities both within and *across* species and to build a transformative politics toward a more democratic, empathetic, and multi-species world (Pellow 2018). These approaches draw on a range of intellectual projects and collective practices that, taken together, recognize the possibility of (and help work toward) 'multispecies justice.'

In terms of justice theory, advocates of multispecies justice highlight the need (i) to deconstruct and decolonize "the hegemony of liberal political discourse"; (ii) to rectify "false assumptions and longstanding misconceptions in justice theory [including] the fictitious idea of human beings as individual, isolated, unattached and unencumbered, and the correlative presumption that more-than-human nature is mere passive background"; and (iii) to rethink the subject of justice so as to account for other beings "with their own radically diverse life projects, capacities, phenomenologies, ways of being, functionings, forms of integrity, and relationalities" (Celermajer et al. 2021, pp. 120, 127). This reflects and connects with the CEJ pillar of indispensability—that the radical embrace of diversity across multiple social and institutional categories strengthens EJ movements by expanding collective understandings of how injustices operate and are experienced. At the same time, these three central foci of multispecies work offer theoretical tools for reconsidering notions of personhood, inter/subjectivity, and responsibility.

Questions around intersubjectivity and responsibility are vital for expanding how we think about different *subjects* of justice claims and processes. One proposal here is to "rephrase the question of harm ontologically"—as a question about which modes of existence (and the suffering they endure) are acknowledged and about whether (and how) those different modes of existence are (or are not) "made to count" as subjects (Reinert 2016, p. 106). This means critically examining who has claimed (and continues to claim) a monopoly over justice (and on what grounds) while also working to challenge such singular claims (Chao et al. 2022). Rephrasing harm ontologically also requires a reassessment (and/or rejection) of the human/nonhuman divide, echoing the account of ontological conflicts, above. One such approach connects justice struggles with a renewed multispecies philosophy that interrogates and qualifies "the broad and seemingly neutral concept of humanity utilized in and around conversations grounded within Western science and academia":

> "The notion of the human, as we know, comes with its own fraught history of exclusion. Who gets to be included in the 'we' employed by scientific, social and political discourses, and who is pushed to the margins and beyond of this seemingly neutral category is a violent story of making invisible, of disenfranchisement, of marginalized bodies and their epistemologies". (Bencke and Bruhn 2022, p. 10)

One branch of influential multispecies work has been developed in the environmental humanities and through post-humanist critical theory. Donna Haraway, for example, argues that there can be no environmental justice without multispecies justice and sees that as a kind of justice which can only be achieved by "nurturing and inventing enduring multispecies (human and nonhuman) kindreds" and by committing to processes of "becoming-with" through "new kinds of person-making"—generating new, experimental ways-of-being and ways-of-relating across categories of species in order to foster other sorts of "selves" (Haraway 2018, p. 102). Other writers and practitioners, including scholars working through multispecies ethnography, build on work that demonstrates how colonial, gendered, and racialized categories of difference exacerbate inequalities within and between species groups in order to go further: to pursue forms of knowledge creation that are "attuned to life's emergence within a shifting assemblage of agentive beings" (Ogden et al. 2013, p. 6). This pushes the question of multispecies justice beyond theoretical concerns about the subject(s) of justice theories by holding all (human and more-than-human) beings and 'selves' within always-emergent webs of relationships, therefore also addressing the kinds of responsibilities that those relationships and interactions entail. The process of identifying inequalities is itself subject to scrutiny, further exploring very different degrees of responsibility for creating (and potentially stopping) environmental harm.

As a challenge to dominant political discourse (and as a more expansive approach that decenters the human in understandings of justice), multispecies justice offers analytical tools that are directly relevant for energy-focused research. For example, in scrutinizing how the construction and operation of energy systems disrupt the freedom and abilities of many different species to lead healthy lives. But the concept is more than merely a diagnostic tool; it also has wide-ranging theoretical, ethical, and legal implications. Theoretically, it emphasizes the contingency—and interdependency—of human and more-than-human worlds, recognizing a wider range of cosmological and ontological possibilities in which the human role is decentered (Gutierrez et al. 2021). Expanded ontological possibilities raise new ethical questions. For example, in some formulations of the concept, this means asking what are considered ethical acts if these relationships of interdependency are to be embodied and enacted within mutually beneficial 'multispecies communities of care' (Fernando 2020). Legally, multispecies justice can support the formal protection of a more comprehensive collection, or community, of beings against environmental harm and injustice. In specific applications and struggles, this is an area where reframing environmental harm in ontological terms can be particularly effective.

Ontological questions are simultaneously questions about relationality. Applied within justice struggles, this is to ask: what are the responsibilities that humans (a small number of who have become the primary agents of planetary destruction) have toward a growing, interconnected, diverse, and emergent gathering of agentive beings? Recentering relationality is vital for pursuing energy transitions as collaborative projects of justice, recognizing the multiple roles that energy relations play in building more just and equitable futures. Multispecies perspectives call for relational ethical approaches that both (i) take seriously multiplying, and changing, conceptions of 'the good' and (ii) recognize that "contesting for better worlds" requires carefully, and repeatedly, crafting shared notions of right and wrong within ongoing processes of cooperation and contestation (van Dooren et al. 2016, p. 16). This builds directly on the pluralist knowledge culture of justice (in Chap. 2, above) that recognizes how the empirical and the normative are inextricably linked (van Gerven 2022). Normative understandings of justice are established through processes of cooperation and contestation, conflict and communication; multispecies justice invites many more beings and experiences to play an active role within those processes.

Contemporary multispecies scholarship is only one source of critical approaches to multispecies justice. As Yazzie (2018) underlines: "it is important to point out that neither Haraway nor any scholar working in post-humanist traditions is offering any new insight into relationality that has not already been expertly theorized and practiced by Diné and other Indigenous peoples since before the advent of American academic institutions" (Yazzie 2018, p. 35). Many underlying concepts—recently rearticulated by some as multispecies justice or interspecies justice—have a long history in critical Indigenous scholarship and a much longer history through diverse Indigenous practices and experiences of relationality, reciprocity, and intersubjectivity (Turner 2006; Figueroa 2013; Whyte 2017). Indigenous feminists, as Yazzie (2018) also highlights, have for a long time emphasized what could be considered a form of multispecies justice—formulating a politics of connection that is rooted in relationality and movement rather than the individualism, hierarchy, heteropatriarchy, and control that typifies settler colonial understandings of time, space, and society (LaDuke 1999; Moreton-Robinson 2002; Goeman 2017). Indigenous feminisms have also applied relationality within a politics that recognizes how language, stories, and acts of imagination offer sources of power to heal and help reformulate inter-species power relations (Goeman and Denetdale 2009).

While acknowledging the profound significance of these insights for fostering planetary health and wellbeing, it is also important to note that the category "species" is itself a social construct based on non-Indigenous perspectives (Nelson 2017). Similar cautions to those discussed above, in Chap. 2, concerning restorative environmental justice, apply here: specifically, the need for academic, non-Indigenous, and institutional practitioners to work against the cooptation and misappropriation of Indigenous knowledges. Expanding and clarifying the many different dimensions and implications of multispecies justice does not mean adopting (or coopting) values, concepts, and ideas from within ongoing Indigenous histories and practices. Rather, just as with restorative environmental justice, the description and pursuit of multispecies justice is best used as a way of building solidarity. Counteracting the concentration of power and associated socio-environmental inequalities requires drawing on multiple fields of activism (Stephens 2020). All struggles for justice are connected and mutually reinforcing. Much direct action is fueled by rebuilding commons and solidarities that "reconnect us to more than ourselves" (Frémeaux and Jordan 2021, p. 42). As Danielle Celermajer emphasizes, there can be no multispecies justice without decolonial, racial, gender, and epistemic justice. Solidarities across struggles for multispecies justice further offer a way of gaining new perspectives on what actions and relations of solidarity might actually involve (especially when those actions seek to build mutual collaborations across multiple categories of social difference).

Restorative EJ involves (i) recognizing the distinct experiences of Indigenous groups and EJ communities while also (re)establishing relationships of respect, care, and reciprocity between everyone engaged in justice struggles; and (ii) at the same time, building collaborative actions to work toward shared futures and against common opponents. So it is with multispecies justice: being attentive to distinct experiences while using this exploratory concept as a basis for exposing and undermining entrenched, dominant perspectives that continue to insist on violent (but ultimately imaginary) divisions between humans and nonhumans. With few exceptions, while those working on the frontlines of Indigenous resistance and those writing from within academia occupy very distinct—and very unequal—positions, what they may share is a focus on "formulating a politics of relational life that can serve as a form of multispecies justice… a critical and necessary framework for liberating all life" from "capitalist, heteropatriarchal, and settler colonial" violence (Yazzie 2018, p. 35). Any moments of overlap that occur offer a point of connection: connecting people interested in exploring the emancipatory and transformative potential of multispecies justice as one of many counter-hegemonic visions for building habitable, and just, planetary futures.

Some struggles for justice led by Indigenous and Native groups have already successfully campaigned for the rights of nonhuman beings or entities to become part of environmental law and policy. The Rights of Nature, as codified in Ecuador's 2008 Constitution, offer protection for the world's ecosystems by drawing directly on the Kichwa idea of *Pachamama*, or Mother Nature. Article #71 in Chapter 7 of the Constitution states that "Nature or Pacha Mama, where life is reproduced and realized, has the right to the integral respect of its existence and the maintenance and regeneration of its life cycles, structure, functions, and evolutionary processes"; Article #72 states the Right of Nature to restoration or reparation (Walsh 2010, p. 18). [1] Highlighting the defense of Indigenous and collective rights in the constitution has been an important strategy in denouncing energy projects that violate those rights, particularly those that exacerbate the dispossession and destruction of Indigenous lands as a result of oil extraction and refining (Acción Ecológica 2018). Still, the (mis)translation and application of Indigenous concepts, perspectives, or values into national political and legal structures is always fraught. Such processes are particularly problematic when—as is practically always the case—those national political and legal systems consistently discriminate against Indigenous people.

Critical EJ work specifically emphasizes the shortcomings of seeking justice through systems "never intended to provide justice for marginalized peoples and nonhuman natures" since states are "social institutions that tend to lean toward practices and relationships that are authoritarian, coercive, racist, patriarchal, exclusionary, militaristic, and anti-ecological" (Pellow 2018, p. 23). The codification of new constitutional or collective rights for Indigenous communities can therefore only ever represent a first step or one component among many within broader justice struggles. Meaningful justice requires addressing the fact that the rights of Indigenous peoples have to be continually defended against state actions that routinely disregard those rights (Baker 2016; Partridge 2016). That said, codified rights can, on occasion, be used strategically (and successfully) by Indigenous groups and environmental movements—as some communities in Ecuador have done in order to enforce limits on particularly destructive or environmentally harmful industries (Acosta and Martínez 2009; Partridge 2017). At the same time, though, such cases also highlight how addressing long-standing inequalities or achieving redistribution requires systemic change beyond the codification of new rights: changes that would involve dismantling the colonial relations which continue to enable some livelihoods (and to respect some peoples' rights) while severely damaging (or curtailing) others (ibid.). Multispecies justice casts these persistent forms of discrimination in a new light by highlighting how ill-equipped legal systems are to provide justice to a diversity of other beings and

---

[1] An alternative translation into English of the complete text of Ecuador's 2008 Constitution (444 articles in total), is available here: https://pdba.georgetown.edu/ Constitutions/Ecuador/english08.html (Accessed: 28 March 2022).

entities. Globally, despite these structural barriers, there are some positive movements in this direction.

Through the Te Urewera Act (2014) and the Te Awa Tupua Act (2017), New Zealand became the first country in the world to grant a legal identity to living lands and to a river (Bargh 2021). In these specific contexts, the former Act was the result of decades of Tūhoe engagement to address breaches by the (British) Crown (and its representatives) of the Te Tiriti o Waitangi (Treaty), securing a "legal identity and protected status for Te Urewera" for "its intrinsic worth, its distinctive natural and cultural values, the integrity of those values, and for its national importance" (Bargh 2021, pp. 426–427). What had been a National Park is now a legal entity. The latter Act stands as a second example of how Māori have successfully negotiated "Māori values and world-views into legislation," settling breaches of Treaty through the Te Awa Tupua (Whanganui River Claims Settlement) Act which affirms Te Awa Tupua as an "indivisible and living whole, comprising the Whanganui River from the mountains to the sea, incorporating its tributaries and all its physical and metaphysical elements" (ibid.). Ecosystem and multispecies *relationships* themselves become subjects of justice (Winter 2021). In these specific contexts, Indigenous struggle had led to some political and legal processes not only recognizing but also incorporating (and expanding to accommodate) Indigenous worldviews, knowledge, and insights.

While not every struggle to legally protect more-than-human beings against environmental harm or to defend the Rights of Nature should be seen as an instance of multispecies justice in action, outcomes including the Te Urewera Act and the Te Awa Tupua Act stand as beacon cases that affirm the potential of radically revising how environmentalisms can operate. Crucially, these moments of victory in the defense of mountains and rivers (and, by extension, in protection of the web of ecosystem relations that support all planetary life) reflect the hard work and hardships of specific justice struggles. These legal Acts came into being as a result of decades of organizing, mobilization, and collective action—pursued through multiple alliances and through multiple forms of engagement with state and community politics. With reference to energy projects, taking seriously the need to protect biodiversity and multiple species—including, but not limited to, endangered species in desert lands identified as having significant solar power potential—not only supports environmental protection but also challenges energy project design to be more imaginative with regard to energy distribution, decentralization, and siting (exploring, for example, the suitability of brownfield sites or other disturbed lands, abandoned mines, or abandoned agricultural land for solar generation) (Mulvaney 2019, p. 117). Multispecies approaches critically reassess how different forms of environmental harm are identified and addressed across multiple scales while also refocusing analytically on the multiple forms of violence that create those harms.

Struggles in defense of the complex relations that connect humans and nature align with struggles against the forces, influences, and institutions that seek to degrade or undermine those connections. We see this in resistance to practices within racial capitalism that impose (and operate through) logics "of discreteness, distinctness, and discontinuity—of discrete identities, distinct territorializations and sovereignties, and discontinuities between the political and the economic, the internal and the external, and the valued and the devalued" (Melamed 2015, p. 79). At issue here is how capital accumulation relies on racialized processes of "spatial and social differentiation"—on the production of social *separateness*—in order to "truncate relationality" and therefore reshape societies such that they more efficiently serve the interests of capital (ibid.). Ruth Wilson Gilmore defines racism as "the state-sanctioned and/or extra-legal production and exploitation of group-differentiated vulnerabilities to premature death, in distinct yet densely interconnected political geographies" (Gilmore 2002, p. 261). Building on this definition, Melamed describes racial capitalism as "a technology of *antirelationality* (a technology for reducing collective life to the relations that sustain neoliberal democratic capitalism)"—as a means of controlling who is allowed to relate and to connect (and controlling the conditions through which those interactions take place) (Melamed 2015, p. 78, original emphasis). Working against such forces, multispecies justice offers tools to more clearly identify—and to deliberately disrupt—the denial of socioecological interrelation through imposed notions of isolation and disconnection. Multispecies justice directly confronts the antirelationality of racial capitalism.

Within any particular place, the specificities of multispecies relationality emerge within particular networks of kin, community, and existence that interact and support each other in that space. At the same time, justice claims and demands—in order to be effective—require broader, external recognition at a number of scales. Multispecies justice as a concept is therefore always operating within a productive tension between the particular and the general: seeking to respect and nurture all life as it takes root and thrives in specific contexts while also calling for justice claims to be heard and upheld by regional, national, and global institutions of power.

Multispecies justice is an adaptive (rather than prescriptive) framework, able to respond to the dynamics of particular political, cultural, ontological, and geographical claims and contexts. Rather than seeking to (mis) appropriate knowledges and practices kept alive by distinct communities or collectives (for example through what might be a limited or compromised incorporation of those practices and related concepts into national legal systems), multispecies justice can instead be used to disrupt ongoing histories of marginalization. That is, by using multispecies justice as a way to open up—and to protect—political, methodological, and relational spaces that allow us to recognize and explore how diverse forms of knowledge can create more holistic, critical, decentered, and grounded approaches to (energy and environmental) justice.

REFLEXIVITY, SOLIDARITY, AND REFUSAL

Renewed practices of reflexivity are paramount for critical energy research. People writing about injustices have a responsibility to commit to more than merely 'questioning one's assumptions' or offering assessments of what a particular field has achieved. A starting point here is to critically examine the political implications of our own research and writing—a fraught and complex process that varies enormously across social, geographical, and political landscapes, but which also centers on some core questions: What relationships have enabled our work? Who has invited us to participate in those relationships? Who are we writing for (and with), and how are we/they positioned (with particular to reference to economic and political power)? Whose values are being upheld? Whose efforts are we aligned with?

Critical energy research seeks to expand the depth and breadth of collaborations that are working to identify, confront, and overcome energy-related injustices. Core goals are: to foreground (and to work in solidarity with) socioecological struggles undertaken by diverse marginalized populations; to acknowledge expertise beyond the academy; and to root scholarship in humility and respect. This orientation reflects—and aims to embrace—the disruptive power of the same social movements that EJ scholarship is learning from. As Benford notes, "the environmental justice movement's power lies in its capacity to disrupt the system rather than seek to reform it" (Benford 2005, p. 52). The 'system' in question—which includes dominant political, economic, and legal structures—is also that which supports institutional knowledge creation.

EJ and Indigenous scholars have addressed questions of reflexivity and respect within research practices in part by reassessing the social, political, and epistemological bases of research relations and by reframing what is meant by collaboration. Instigating restorative research relations rooted in solidarity requires a critical rejection of dominant, extractive modes of inquiry. These are co-constitutive, iterative processes that reinforce each other and evolve over time: exploring new ways of doing research that create new insights and ideas, which inform revitalized research collaborations and reassessments of reflexivity, in an ongoing cycle which repeats and continues to productively challenge the bases and objectives of research practices themselves. Drawing on anarcho-feminist theory and praxis—particularly the work of Victoria Aldunate Morales and the Mujeres Creando Comunidad collective in Bolivia—Macarena Gómez-Barris explores 'modes of seeing' that are embodied, non-essentialist, and in opposition to the "singular, patriarchal, and hierarchical organizing vision of gendered capitalist economies," envisioning instead "theories, activities, and solidarities that the state cannot reduce, or that extractive capitalist economies cannot fully capture" (Gómez-Barris 2017, p. 124). Co-creating restorative research relations in these ways requires a critical reassessment of our own principles and ideas, including how solidarity itself is articulated and practiced.

Scrutinizing the ways in which racism is a barrier to solidarity between women, bell hooks specifically challenges commonplace understandings of solidarity as insufficient for building a more inclusive, liberatory movement that "challenges rather than perpetuates domination" and which enables stronger political cooperation:

> "Solidarity is not the same as support. To experience solidarity, we must have a community of interests, shared beliefs and goals around which to unite, to build Sisterhood. Support can be occasional. It can be given and just as easily withdrawn. Solidarity requires sustained, ongoing commitment". (hooks 1986, p. 138)

Reflecting on her work as a nonnative activist allying with Indigenous communities to fight against the ongoing colonization of Indigenous lands and against the ongoing assimilation of Indigenous world-views and practices, Harsha Walia (2012) cites the same hooks text and applies it to her own attempts to act in meaningful solidarity with Indigenous struggles. This work—striving toward decolonization—requires non-natives to not only recognize our role in perpetuating colonialism (even, or especially, within our solidarity efforts) but also to actively counter those effects by pursuing solidarity through actual, sustained, informed relationships (rather than as an abstraction or ideal) (Walia 2012, p. 359). Pursuing this line of thought, Walia adds that decolonization might even require getting rid of terminology such as "solidarity" since it is a concept that "implicitly signifies the 'otherness' of those with whom one is in relation" and often "reduces our ability to be in kinship with one another" (ibid.). The goal remains instead to challenge "a dehumanizing social organization that perpetuates our isolation from each other and normalizes a lack of responsibility to one another and the Earth," thus rejecting an assumed or simple "unity across our differences, in particular those that are rooted in systems of power and privilege" (Walia 2012, pp. 359–360). This means creating "a radical terrain of struggle where our common visions for justice do not erase our different social locations" and where "our differing identities do not prevent us from walking together toward transformation and mutual respect" (ibid.).

Solidarity—if it is to be revived and reimagined—has to be part of unfolding active, relational, and mutual processes. This is fundamental within decolonization efforts and in contexts where researchers seek to support Indigenous struggles for justice. But the same demands apply wherever justice-focused academic work seeks to write about and to operate in solidarity with engaged movements and communities. Solidarity also means starting from a point of engagement with those who are already tackling the issues and injustices that scholarly work also seeks to address—recognizing (and learning humbly and respectfully from) those already-underway processes while also rejecting the customary academic tendency to claim 'firstness' (especially when doing so obscures work done previously outside of academia).

Emphasizing that these processes are fraught, complex, and context-dependent—and finding ways to better articulate and respond to these dynamics—means recognizing that such work is never finished. Jaskiran Dhillon underlines how, like all other forms of collective social and political work, practices of research and solidarity are fundamentally "bound up with how we develop a sense of connection to one another" and so "research and advocacy for social transformation is far from a benign enterprise; it is always about power" (Dhillon 2019, pp. 41–42).

Scrutinizing how 'solidarity' is conceptualized and put into practice is one dimension of renewing reflexivity. Another dimension concerns how 'allyship' is enacted and expressed. A number of scholars draw on a text by Indigenous Action (2014) in order to critique widespread notions of allyship and alliance-making within research processes. The piece—"Accomplices Not Allies"—addresses tensions surrounding different forms of solidarity/support work and has particular points of advice for those working in the academic sector:

> "Allyship is the corruption of radical spirit and imagination, it's the dead end of decolonization... What is not understood is that decolonization is a threat to the very existence of settler 'allies'... Although sometimes directly from communities in struggle, intellectuals' and academics' [role] in struggle can be extremely patronizing. In many cases the academic maintains institutional power above the knowledge and skill base of the community/ies in struggle... An accomplice as academic would seek ways to leverage resources and material support and/or betray their institution to further liberation struggles. An intellectual accomplice would strategize with, not for and not be afraid to pick up a hammer". (Indigenous Action 2014, n.)

This intervention also queries an intellectual fixation on "un-learning oppression" by asking whether we should desire to "merely 'unlearn' oppression" or instead fashion weapons to fight it directly: "Direct action is really the best and may be the only way to learn what it is to be an accomplice. We're in a fight, so be ready for confrontation and consequence" (Indigenous Action 2014, n.). For those of us in such a position, the work of unlearning *and* undoing our own settler-colonial mentality has to be continuous (Temper 2019, p. 108, emphasis added). Even if processes of unlearning are understood to be a vital step toward meaningful engagement, the Accomplices Not Allies intervention is a clear reminder that those critical processes remain incomplete without also building on them by taking consequential action.

These ideas have rightfully resonated across disciplines and projects, urging practitioners to reframe, reimagine, and to re-engage with collective research practices. Responding to the latest (and predictable) wave of open political commitment to white nationalism and authoritarianism among those in the highest political offices of the US, anthropologist Mark Schuller calls for people to recognize (and live up to) the responsibilities demanded of everyone in positions of privilege: "to move beyond being *allies*—where privilege remains intact—to becoming *accomplices,* acknowledging self-interest and putting one's own body on the line, disrupting and dismantling privilege" (Schuller 2020, p. 176, original emphasis). Dhillon makes a parallel argument (substituting 'comrades' for 'accomplices') and illustrates how substantial reflexivity requires practitioners to develop (and make explicit) a critical, institutional, and personal self-awareness:

> "I use the term 'comrade' rather than 'ally' to signal urgency around the politicization of solidarity work. Benign notions of allyship, or solidarity with no teeth, has been rightfully critiqued for reinscribing colonial relations of domination, and doing little to interrupt or dismantle white settler power and broader colonial structures upholding white supremacy. Thinking of oneself as a political comrade demands, first and foremost, that you know who you are, that you have a deep understanding of what you have inherited by virtue of your social history and political standing". (Dhillon 2019, p. 43)

Another way to renew the meaning (and practice) of reflexivity within critical energy research is to engage in a 'politics of refusal.' There are echoes here of Simpson's (2014) work on refusal in the case of Indigenous political action (discussed above in Chap. 1: Settler Colonialism and Sovereignty); the focus now, though, is on other institutional contexts. Specifically, this is a refusal of practices that "reproduce the settler, imperial, and carceral logics of mainstream environmentalism or the corporate university" (Di Chiro 2020, p. 322). That is, to refuse practices of 'epistemic extractivism' and instead to replace them with processes of acknowledging and deepening acts of reciprocity (Grosfoguel 2019). Refusal here is as an ethical position that demands practitioners find new ways to reject the structures of inequality that their (professional or social) roles build upon and replicate.

Further detailing this approach—and explicitly seeking accomplices not allies—Grande (2018) draws on Indigenous practices of refusal to outline how a broader politics of refusal takes shape through three forms of commitment: (i) to *collectivity,* by rejecting the "individualist promise project" of settler states and their attendant institutions; (ii) to *reciprocity,* by being answerable to "those communities we claim as our own and those we claim to serve" and pursuing a form of accountability that demands we "engage, extend, trouble, speak back to, and intensify our words and deeds"; and (iii) to *mutuality,* by radically reaffirming connection to land

and to "the arc of inter-generational resurgence and transformation" that supports respectful intersubjectivity (Grande 2018, p. 61, original emphasis). As with renewed approaches to solidarity, these are active, relational, and always-unfinished processes. Citing Garland (2013, p. 375), Grande specifies that "refusal should not be confused with 'passive withdrawal or retreat' but rather be understood as an active instantiation of 'a radically different mode-of-being and mode-of-doing'" (Grande 2018, p. 58). Rooting critical academic reflexivity in a politics of refusal—rather than (only) in disciplinary debates, reviews, or expectations—offers ideas and generative practices for reformulating energy-focused research as a mode of meaningful intervention within justice-focused struggles.

While engagement with these different forms of refusal offer important guidelines for recalibrating the political potential of critical energy research, it is necessary to recognize that such processes of engagement are never straightforward. As with many modes of resistance, acts of refusal may generate new ideals and practices of equality but they may also provoke, participate in, or leave untouched the forms of hierarchy that those involved would typically oppose (Graeber 2013). Crucially, though, a politics of refusal serves to connect energy research ever more closely with some of the core praxis-oriented EJ principles described in Chap. 2: pursuing greater autonomy from exclusive institutions; refusing practices that replicate violent, hierarchical relationships; and creating—and holding open—political spaces in support of careful and committed forms of collaboration.

CRITICAL ENERGY RESEARCH

This book has mapped out key issues for a critical energy research agenda by highlighting EJ scholars and activists whose work expands how we think about and address energy-related injustices. Critical energy research connects the study of energy relations and infrastructures with these multiple EJ perspectives, recognizing energy as an inseparable aspect of broader political, ecological, economic, and (re)productive relations. Using these approaches, critical energy research analyzes links between the operation of energy systems and environmental degradation, racism, violence against Indigenous bodies, gendered oppression, and multiple forms of injustice and harm enacted against diverse beings who are rendered 'other' within dominant modes of societal organization. Repoliticizing energy histories and potential futures plays a central role in these analyses, countering the anti-politics of much energy research and elevating work already being done toward these ends, especially in nonformal, cooperative, community-based, or activist-led contexts. This means rejecting the assumption that those who currently own, operate, govern, and profit from energy systems should continue to do so; tracing how political-economic histories have produced (and continue to replicate) current conditions of injustice; and rejecting the idea that social inequalities and processes of marginalization are merely technical challenges. Critical energy research draws on an evolving range of theoretical, methodological, and political ideas to develop an expansive approach to energy research. These ideas, gathered below, focus on the specificities of energy relations (and the materials, processes, and socioecological connections that constitute them) as a way of reassessing broader processes of social, political, and environmental change:

A global movement of movements: Critical energy research foregrounds the role of EJ communities, Indigenous groups, and diverse social movements, continuing the exchange of concepts and strategies between activists and academics that has long been central to EJ (Martinez-Alier et al. 2014). The contemporary scale and expansion of energy systems (and growing global energy demand) are primary drivers of both climate change and the eradication of increasing amounts of global biodiversity. Confronting these trends, critical energy research connects with movements for climate justice that expose links between energy systems and underlying causes of climate injustices (including dispossession, exploitation, and industrial expansion). Critical energy research also maintains that any notion of 'energy justice' needs to be rooted in diverse EJ and Indigenous activism, focused on systemic injustices, and working to counter the uncritical appropriation of justice discourse by state, corporate, and other institutional actors. Drawing on the work of prefigurative, autonomous movements, this means exploring energy justice as an unfolding process: a spectrum of political engagements that identify both how energy relations impinge upon broader struggles for transformative social change and, conversely, how energy relations can be reimagined in order to support those same struggles.

Systems of production and extractivism: Justice-oriented energy research maintains a critical focus on how ongoing processes of settler colonialism and multiple forms of exploitation interact and underpin energy-intensive systems of capitalist production. Widespread, implicit acceptance of the idea that energy is the capacity to do commodifiable work perpetuates the production of social and environmental injustices, in part by denying how economic activity fundamentally depends on non-mechanical work and on ecosystems that are healthy enough to regenerate. Diverse bodies and beings are systematically devalued under patriarchal racial capitalism, evident in dominant modes of organizing productive relations that depend upon and enforce discrimination across multiple social categories of difference, including race, gender, class, caste, sexuality, ethnicity, ability, and species. Exposing these dominant trends, critical energy research connects with work in ecofeminism and other fields that study the exploitation of people and Nature as processes that are inseparable both from each other and from economic activities committed solely or primarily to growth. Critical energy research also connects with the study of extractivism and its effects, including how energy relations contribute to unequal global development, to an antidemocratic concentration of power and wealth among transnational corporate actors, and to economic processes based upon the privatization of profits and the socialization of costs.

Transitions as collaborative projects of justice: Beyond standard framings of the energy transition in terms of crisis (which are not universally understood or experienced), critical energy research studies all transition practices and discourse in relation to their capacity to facilitate transitions not just to different energy systems but also to more inclusive, equitable, and reciprocal ways of living and relating. These approaches recognize that resolving energy-related injustices implies the need for structural transformations of productive and political systems, beyond the application of Just Transition policies. Strategies for implementing or initiating transformative change include planned and equitable degrowth—a direct response to the fact that increasing global energy use consistently exacerbates the negative social and environmental effects of racialized systems of value production. Transitions as collaborative projects of justice respect and protect Indigenous sovereignty, acknowledge that recognition and participation might need to be reframed (or rejected) in order to realize justice claims, and scrutinize whether extant political and judicial systems are equipped to fulfil justice demands made by marginalized communities.

Restorative environmental justice: Critical energy research focuses on justice as the restoration of mutually supportive relationships between beings of all kinds. This directly connects with the pillar of *indispensability* within Critical Environmental Justice: recognizing the necessary roles of all humans and more-than-human actors in building our shared futures (and in working against practices that render certain populations expendable). Restorative environmental justice draws on multispecies justice in order to directly confront the antirelationality of racial capitalism. It also pursues multispecies justice as a way to open up (and to protect) spaces for the work of resisting, reorganizing, and repairing destructive socioecological relations.

Renewed reflexivity. Reflexivity demands that we, as researchers, subject our own analyses, research relations, and notions of solidarity to ongoing political scrutiny. Renewed reflexivity requires recognizing how processes including research collaborations, ethnography, allyship, and activist strategy-building are themselves sites for enacting justice (Forbes et al. 2021; Gutierrez et al. 2021). At the same time, reflexivity also requires recognizing how such sites and processes are always fraught with the risk that practitioners will replicate the asymmetries and inequalities that justice-oriented work nominally seeks to dismantle. Scholars need to think not only critically but also humbly about their role in working toward energy and environmental justice (Grosse 2022). Critical energy research draws on three different fields to deepen these practices: (i) approaches to solidarity rooted in restorative research relations and a critical rejection of dominant, extractive modes of inquiry, aiming to work as an accomplice rather than an ally; (ii) a politics of refusal, which is an explicit rejection of practices that rely on harmful hierarchical relations; (iii) acknowledging multispecies relations within reflexivity and holding all human and more-than-human beings within emergent webs of relationships; this also means incorporating consideration of the responsibilities that those relationships entail within research processes.

Injustices are painfully clear to anyone who has to suffer the toxification of the land they live on, of the air they breathe, or of the water that flows through their bodies and their world. While some energy researchers come directly from communities in struggle, not all (perhaps relatively few) have these experiences. Nor is it always clear how our work as researchers can resolve energy and environmental injustices. Nevertheless, there remain spaces for researchers to take up political, analytical, documentary, and network-building roles; to expose the racism, colonialism, and other forms of violence routinely carried out by energy corporations, states, and other powerful or institutional actors; to stand as witnesses; to share knowledge across sites of struggle and conflict; to leverage positions of privilege and redirect to others the resources that privileges bring; to stand back, to decenter our role, to discomfort ourselves and reassess the ways in which we are taking up space; to continue developing careful and committed research practices while learning, sharing, and engaging with the global movement of EJ movements. These are some of the core goals of critical energy research; points of departure for continuing collective work. They are goals to guide our actions in restoring relations of care, respect, and reciprocity within growing global movements for peace and justice.

REFERENCES

Acción Ecológica. 2018. *Más de 10 razones para no invertir en la Refinerías del Pacífico*. Quito: Acción Ecológica Opina.

Acosta, A., and E. Martínez, eds. 2009. *Plurinacionalidad: Democracia en la diversidad*. Quito: Abya-Yala.

Adamson, J. 2011. Medicine Food: Critical Environmental Justice Studies, Native North American Literature, and the Movement for Food Sovereignty. *Environmental Justice* 4 (4): 213–219.

AI. 2017. *Investigate Shell for Complicity in Murder, Rape and Torture*. London: Amnesty International.

Avila, S., Deniau, Y., Sorman, A.H., and McCarthy, J. 2021. (Counter)mapping Renewables: Space, Justice, and Politics of Wind and Solar Power in Mexico. *Environment and Planning E: Nature and Space*. OnlineFirst.

Baker, S.H. 2016. Mexican Energy Reform, Climate Change, and Energy Justice in Indigenous Communities. *Natural Resources Journal* 56 (2): 369–290.

Bakker, K. 2012. Water: Political, Biopolitical, Material. *Social Studies of Science* 42 (4): 616–623.

Bargh, M. 2021. Diverse Indigenous Environmental Identities: Māori Resource Management Innovations. In *Routledge Handbook of Critical Indigenous Studies*, ed. B. Hokowhitu, A. Moreton-Robinson, L. Tuhiwai-Smith, C. Andersen, and S. Larkin, 420–430. New York: Routledge.

Bencke, I., and J. Bruhn. 2022. Introduction. In *Multispecies Storytelling in Intermedial Practices*, ed. I. Bencke and J. Bruhn, 9–20. Santa Barbara: punctum books.

Benford, R. 2005. The Half-Life of the Environmental Justice Frame: Innovation, Diffusion, and Stagnation. In *Power, Justice, and the Environment: A Critical Appraisal of the Environmental Justice Movement*, ed. D.N. Pellow and R.J. Brulle, 37–53. The MIT Press.

Bickerstaff, K., G. Walker, and H. Bulkeley, eds. 2013. *Energy Justice in a Changing Climate: Social Equity and Low-Carbon Energy*. London: Zed Books.

Blaser, M. 2009. Political Ontology: Cultural Studies Without 'cultures'? *Cultural Studies* 23 (5–6): 873–896.

Boelens, R., and M. Seemann. 2014. Forced Engagements: Water Security and Local Rights Formalization in Yanque, Colca Valley, Peru. *Human Organization* 73 (1): 1–12.

Bruno, K., J. Karliner, and C. Brotsky. 1999. *Greenhouse Gangsters vs. Climate Justice*. San Francisco: Transnational Resource and Action Center (TRAC).

Building Bridges Collective. 2010. *Space For Movement? Reflections from Bolivia on Climate Justice, Social Movements and the State*. Leeds: Footprint Workers Co-op.

Bullard, R.D. 1990. *Dumping in Dixie: Race, Class and Environmental Quality*. Boulder: Westview Press.

———. 1996. Environmental Justice: It's More Than Waste Facility Siting. *Social Science Quarterly* 77 (3): 493–499.

de la Cadena, M., and M. Blaser, eds. 2018. *A World of Many Worlds*. Durham: Duke University Press.

Castán Broto, V., I. Baptista, J. Kirshner, S. Smith, and S. Neves Alves. 2018. Energy Justice and Sustainability Transitions in Mozambique. *Applied Energy* 228: 645–655.

Celermajer, D., D. Schlosberg, L. Rickards, M. Stewart-Harawira, M. Thaler, P. Tschakert, B. Verlie, and C. Winter. 2021. Multispecies Justice: Theories, Challenges, and a Research Agenda for Environmental Politics. *Environmental Politics* 30 (1–2): 119–140.

Chao, S., K. Bolender, and E. Kirksey, eds. 2022. *The Promise of Multispecies Justice*. Durham, NC: Duke University Press.

Ciplet, D. 2021. From Energy Privilege to Energy Justice: A Framework for Embedded Sustainable Development. *Energy Research & Social Science* 75: 101996.

Damgaard, C.S., D. McCauley, and L. Reid. 2022. Towards Energy Care Ethics: Exploring Ethical Implications of Relationality Within Energy Systems in Transition. *Energy Research & Social Science* 84: 102356.

Day, R. 2020. Energy Justice. In *Environmental Justice: Key Issues*, ed. B. Coolsaet, 161–175. Oxford: Routledge.

Dhillon, J. 2019. Notes on Becoming a Comrade: Indigenous Women, Leadership, and Movement(s) for Decolonization. *American Indian Culture and Research Journal* 43 (3): 41–54.

Di Chiro, G. 2020. Mobilizing 'intersectionality' in Environmental Justice Research and Action in a Time of Crisis. In *Environmental Justice: Key Issues*, ed. B. Coolsaet, 316–333. Oxford: Routledge.

van Dooren, T., E. Kirksey, and U. Münster. 2016. Multispecies Studies: Cultivating Arts of Attentiveness. *Environmental Humanities* 8 (1): 1–23.

EJNA. 2009. *Energy Justice in Native America: A Policy Paper for Consideration by the Obama Administration and the 111th Congress*. Honor the Earth; Intertribal Council On Utility Policy; International Indian Treaty Council; Indigenous Environmental Network.

Energy Justice Network. n.d. About Energy Justice Network [Online]. http://www.energyjustice.net/about. Accessed 19 October 2018.

Escobar, A. 2020. *Pluriversal Politics: The Real and the Possible.* Durham, NC: Duke University Press.

Estes, N. 2019. *Our History Is the Future: Standing Rock versus the Dakota Access Pipeline, and the Long Tradition of Indigenous Resistance.* London: Verso.

Featherstone, D. 2005. Towards the Relational Construction of Militant Particularisms: Or Why the Geographies of Past Struggles Matter for Resistance to Neoliberal Globalisation. *Antipode* 37 (2): 250–271.

Fernando, J. 2020. From the Virocene to the Lovecene Epoch: Multispecies Justice as Critical Praxis for Virocene Disruptions and Vulnerabilities. *Journal of Political Ecology* 27 (1): 685–731.

Figueroa, I. 2006. Indigenous Peoples Versus Oil Companies: Constitutional Control Within Resistance. *Sur: International Journal on Human Rights* 4: 51–80.

Figueroa, R.M. 2013. Risking Recognition: New Assessment Strategies for Environmental Justice and American Indian Communities. *American Philosophical Association Newsletter on Indigenous Philosophy* 12 (2): 4–10.

Finley-Brook, M., T.L. Williams, J.A. Caron-Sheppard, and M.K. Jaromin. 2018. Critical Energy Justice in US Natural Gas Infrastructuring. *Energy Research & Social Science* 41: 176–190.

Forbes, R., S. Wochele, K. Peterson, and A. Craggs. 2021. Environmental Justice and Black Lives Matter: Critical Reflection and Advocacy for Social Work in the United States. *Environmental Justice* 14 (6): 404–410.

Fraser, N. 2000. Rethinking Recognition. *New Left Review* 3: 107–120.

Frémeaux, I., and J. Jordan. 2021. *We Are 'nature' Defending Itself: Entangling Art, Activism and Autonomous Zones.* London: Pluto Press.

Gaard, G.C. 2017. *Critical Ecofeminism.* Lanham: Lexington Books.

Galvin, R. 2020. "Let justice roll down like waters": Reconnecting Energy Justice to Its Roots in the Civil Rights Movement. *Energy Research & Social Science* 62: 101385.

Garland, C. 2013. Negating That Which Negates Us: Marcuse, Critical Theory, and the New Politics of Refusal. *Radical Philosophy Review* 16 (1): 375–385.

Gedicks, A. 1993. *The New Resource Wars: Native and Environmental Struggles Against Multinational Corporations.* Cambridge: South End Press.

van Gerven, J.P. 2022. *The Anti-Nuclear Power Movement and Discourses of Energy Justice.* Lanham: Lexington Books.

Gilmore, R.W. 2002. Race and Globalization. In *Geographies of Global Change: Remapping the World*, ed. R.J. Johnston, P.J. Taylor, and M. Watts. New York: Wiley-Blackwell.

Goeman, M. 2017. Ongoing Storms and Struggles: Gendered Violence and Resource Exploitation. In *Critically Sovereign: Indigenous Gender, Sexuality, and Feminist Studies*, ed. J. Barker, 99–126. Durham: Duke University Press.

Goeman, M., and J. Denetdale. 2009. Native Feminisms: Legacies, Interventions, and Indigenous Sovereignties. *Wicazo Sa Review* 24 (2): 9–13.

Goldtooth, T. 1995. Indigenous Nations: Summary of Sovereignty and Its Implications for Environmental Protection. In *Environmental Justice: Issues, Policies, and Solutions*, ed. B. Bryant, 138–148. Washington, DC: Island Press.

Goldtooth, T.B., and M. Awanyanka. 2010. The State of Indigenous America Series: Earth Mother, Piñons, and Apple Pie. *Wicazo Sa Review* 25 (2): 11–28.

Gómez-Barris, M. 2017. *The Extractive Zone: Social Ecologies and Decolonial Perspectives*. Durham, NC: Duke University Press.

Graeber, D. 2013. Culture as Creative Refusal. *The Cambridge Journal of Anthropology* 31 (2): 1–19.

Grande, S. 2018. Refusing the University. In *Toward What Justice?: Describing Diverse Dreams of Justice in Education*, ed. E. Tuck and K.W. Yang, 47–65. New York: Routledge.

Grosfoguel, R. 2019. Epistemic Extractivism: A Dialogue with Alberto Acosta, Leanne Betasamosake Simpson, and Silvia Rivera Cusicanqui. In *Knowledges Born in the Struggle: Constructing the Epistemologies of the Global South*, ed. B.S. de Santos and M.P. Meneses, 203–218. New York: Routledge.

Grosse, C. 2022. *Working Across Lines: Resisting Extreme Energy Extraction*. Oakland: University of California Press.

Gruen, L. 2014. Facing Death and Practicing Grief. In *Ecofeminism: Feminist Intersections with Other Animals and the Earth*, ed. C. Adams and L. Gruen, 127–141. New York: Bloomsbury.

Gutierrez, G.M., D.E. Powell, and T.L. Pendergrast. 2021. The Double Force of Vulnerability: Ethnography and Environmental Justice. *Environment and Society* 12 (1): 66–86.

Haraway, D. 2018. Staying with the Trouble for Multispecies Environmental Justice. *Dialogues in Human Geography* 8 (1): 102–105.

Harvey, D. 1996. *Justice, Nature, and the Geography of Difference*. Cambridge: Blackwell Publishers.

Healy, N., and J. Barry. 2017. Politicizing Energy Justice and Energy System Transitions: Fossil Fuel Divestment and a "just transition". *Energy Policy* 108: 451–459.

Hernández, D. 2015. Sacrifice Along the Energy Continuum: A Call for Energy Justice. *Environmental Justice* 8 (4): 151–156.

Hernández, D., L. Yoon, and N. Simcock. 2022. Basing "Energy Justice" on Clear Terms: Assessing Key Terminology in Pursuit of Energy Justice. *Environmental Justice* 15 (3): 127–138.

Hess, C.E.E., and W.C. Ribeiro. 2016. Energy and Environmental Justice: Closing the Gap. *Environmental Justice* 9 (5): 153–158.

Hidalgo, J.P., R. Boelens, and J. Vos. 2017. De-colonizing Water. Dispossession, Water Insecurity, and Indigenous Claims for Resources, Authority, and Territory. *Water History* 9 (1): 67–85.

Holifield, R., M. Porter, and G. Walker. 2010. Introduction: Spaces of Environmental Justice – Frameworks for Critical Engagement. In *Spaces of Environmental Justice*, ed. R. Holifield, M. Porter, and G. Walker, 1–22. Chichester: Wiley-Blackwell.

hooks, bell. 1986. Sisterhood: Political Solidarity Between Women. *Feminist Review* 23: 125–138.

Indigenous Action. 2014. Accomplices Not Allies: Abolishing the Ally Industrial Complex. *Indigenous Action*, May 2.

Jasanoff, S. 2018. Just Transitions: A Humble Approach to Global Energy Futures. *Energy Research & Social Science* 35: 11–14.

Jenkins, K. 2018. Setting Energy Justice Apart from the Crowd: Lessons from Environmental and Climate Justice. *Energy Research & Social Science* 39: 117–121.

Jenkins, K.E.H., B.K. Sovacool, N. Mouter, N. Hacking, M.-K. Burns, and D. McCauley. 2021. The Methodologies, Geographies, and Technologies of Energy Justice: A Systematic and Comprehensive Review. *Environmental Research Letters* 16: 043009.

Joy, K.J., S. Kulkarni, D. Roth, and M. Zwarteveen. 2014. Re-politicising Water Governance: Exploring Water Re-allocations in Terms of Justice. *Local Environment* 19 (9): 954–973.

LaDuke, W. 1981. Red Land and Uranium Mining: How the Search for Energy Is Endangering Indian Tribal Lands. *Radcliffe Quarterly*, December, 15–17.

———. 1999. *All Our Relations: Native Struggles for Land and Life*. Cambridge; Minneapolis: South End Press/Honor the Earth.

Lee, J., and J. Byrne. 2019. Expanding the Conceptual and Analytical Basis of Energy Justice: Beyond the Three-Tenet Framework. *Frontiers in Energy Research* 7: 99.

Martinez-Alier, J. 2002. *The Environmentalism of the Poor: A Study of Ecological Conflicts and Valuation*. Cheltenham: Edward Elgar.

Martinez-Alier, J., I. Anguelovski, P. Bond, D.D. Bene, F. Demaria, J.-F. Gerber, L. Greyl, W. Haas, H. Healy, V. Marín-Burgos, G. Ojo, M. Porto, L. Rijnhout, B. Rodríguez-Labajos, J. Spangenberg, L. Temper, R. Warlenius, and I. Yánez. 2014. Between Activism and Science: Grassroots Concepts for Sustainability Coined by Environmental Justice Organizations. *Journal of Political Ecology* 21: 19–60.

Martinez-Alier, J., L. Temper, D. Del Bene, and A. Scheidel. 2016. Is There a Global Environmental Justice Movement? *The Journal of Peasant Studies* 43 (3): 731–755.

Massey, D. 1999. *Power-geometries and the Politics of Space-time*. Department of Geography: University of Heidelberg.

Melamed, J. 2015. Racial Capitalism. *Critical Ethnic Studies* 1 (1): 76–85.

Montoya, T. 2016. Violence on the Ground, Violence Below the Ground. *Hot Spots, Cultural Anthropology Website*.

Moreton-Robinson, A. 2002. *Talkin' Up to the White Woman: Indigenous Women and Feminism*. St. Lucia: University of Queensland Press.

Mulvaney, D. 2019. *Solar Power: Innovation, Sustainability, and Environmental Justice*. Oakland: University of California Press.

Nader, L., ed. 2010. *The Energy Reader*. Chichester: Wiley-Blackwell.

Nelson, M.K. 2017. Getting Dirty: The Eco-Eroticism of Women in Indigenous Oral Literatures. In *Critically Sovereign: Indigenous Gender, Sexuality, and Feminist Studies*, ed. J. Barker, 229–260. Durham, NC: Duke University Press.

Newell, P. 2021. Race and the Politics of Energy Transitions. *Energy Research & Social Science* 71: 101839.

Newell, P., and D. Mulvaney. 2013. The Political Economy of the 'just transition'. *The Geographical Journal* 179 (2): 132–140.

Ngwakwe, C.C. 2021. Niger Delta Oil Spill Case Against the Shell Company at the Hague: The Future of Corporate Environmental Responsibility. *Acta Universitatis Danubius Juridica* 17 (2): 27–39.

NNAF. 2019. *People of Asia Say No to Nuclear Power: No Nukes Asia Forum, Japan*. Shahpur Jat: Yoda Press.

Ogden, L.A., B. Hall, and K. Tanita. 2013. Animals, Plants, People, and Things: A Review of Multispecies Ethnography. *Environment and Society* 4 (1).

Pallares, A. 2002. *From Peasant Struggles to Indian Resistance: The Ecuadorian Andes in the Late Twentieth Century*. Norman: University of Oklahoma Press.

Partridge, T. 2015. Recoupling Groups Who Resist: Dimensions of Difference, Opposition and Affirmation. *Journal of Resistance Studies* 1 (2): 12–50.

———. 2016. Rural intersections: Resource Marginalisation and the "non-Indian problem" in Highland Ecuador. *Journal of Rural Studies* 47: 337–349.

———. 2017. Resisting Ruination: Resource Sovereignties and Socioecological Struggles in Cotopaxi, Ecuador. *Journal of Political Ecology* 24: 763–776.

———. 2020. "Power farmers" in North India and New Energy Producers Around the World: Three Critical Fields for Multiscalar Research. *Energy Research & Social Science* 69: 101575.

Pellegrini-Masini, G., A. Pirni, and S. Maran. 2020. Energy Justice Revisited: A Critical Review on the Philosophical and Political Origins of Equality. *Energy Research & Social Science* 59: 101310.

Pellow, D.N. 2014. *Total Liberation: The Power and Promise of Animal Rights and the Radical Earth Movement*. Minneapolis: University of Minnesota Press.

———. 2018. *What Is Critical Environmental Justice?* Cambridge: Polity Press.

Pellow, D.N., and R.J. Brulle. 2005. Power, Justice, and the Environment: Toward Critical Environmental Justice Studies. In *Power, Justice, and the Environment: A Critical Appraisal of the Environmental Justice Movement*, ed. D.N. Pellow and R.J. Brulle, 1–19. The MIT Press.

Perreault, T., R. Boelens, and J. Vos. 2018. Introduction: Re-politicizing Water Allocation. In *Water Justice*, ed. R. Boelens, T. Perreault, and J. Vos, 34–42. Cambridge University Press.

126   T. PARTRIDGE

POCELS. 1991. *The Principles of Environmental Justice, Drafted and Adopted by Delegates to the First National People of Color Environmental Leadership Summit.* October 24–27: Washington, DC.

———. 2002. *Principles of Working Together (adopted at the Second People of Color Environmental Leadership Summit / Summit II).* October 23–27: Washington, DC.

Ranco, D.J. 2008. The Trust Responsibility and Limited Sovereignty: What Can Environmental Justice Groups Learn from Indian Nations? *Society & Natural Resources* 21 (4): 354–362.

Reinert, H. 2016. About a Stone: Some Notes on Geologic Conviviality. *Environmental Humanities* 8 (1): 95–117.

Robertson, A. 2021. *Linked Fates: How California's Oil Imports Affect the Future of the Amazon Rainforest.* San Francisco: Stand.earth / Amazon Watch.

Roth, D., M. Zwarteveen, K. Joy, and S. Kulkarni. 2014. Water Rights, Conflicts, and Justice in South Asia. *Local Environment* 19 (9): 947–953.

Ryder, S., K. Powlen, M. Laituri, S. Malin, J. Sbicca, and D. Stevis, eds. 2021. *Environmental Justice in the Anthropocene: From (Un)Just Presents to Just Futures.* London: Routledge.

Saro-Wiwa, K. 1995. *A Month and a Day: A Detention Diary.* London: Penguin.

Sawyer, S. 2004. *Crude Chronicles: Indigenous Politics, Multinational Oil, and Neoliberalism in Ecuador.* Durham, NC: Duke University Press.

Schlosberg, D. 2007. *Defining Environmental Justice: Theories, Movements, and Nature.* Oxford University Press.

———. 2013. Theorising Environmental Justice: The Expanding Sphere of a Discourse. *Environmental Politics* 22 (1): 37–55.

Schuller, M. 2020. Challenges of "Communiversity" Organizing in Trumplandia. In *Anthropology and Activism: New Contexts, New Conversations,* ed. A.J. Willow and K.A. Yotebieng, 175–189. New York: Routledge.

Simpson, A. 2014. *Mohawk Interruptus: Political Life across the Borders of Settler States.* Durham: Duke University Press.

Skelton, R., and V. Miller. 2016. *The Environmental Justice Movement.* New York: NRDC.

Smith, S.L., and B. Frehner, eds. 2010. *Indians and Energy: Exploitation and Opportunity in the American Southwest.* Santa Fe: School for Advanced Research Press.

Sovacool, B.K., R.J. Heffron, D. McCauley, and A. Goldthau. 2016. Energy Decisions Reframed as Justice and Ethical Concerns. *Nature Energy* 1 (5): 16024.

Stephens, J.C. 2020. *Diversifying Power: Why We Need Antiracist, Feminist Leadership on Climate and Energy.* Washington, DC: Island Press.

Szolucha, A. 2018. Introduction: Conceptualising Energy Impacts and Contested Energy Futures. In *Energy, Resource Extraction and Society: Impacts and Contested Futures,* ed. A. Szolucha. Abingdon: Routledge.

Temper, L. 2013. *Crude Justice & Ecocide in the Niger Delta*. Barcelona: EJOLT.
———. 2019. Blocking Pipelines, Unsettling Environmental Justice: From Rights of Nature to Responsibility to Territory. *Local Environment* 24 (2): 94–112.
Temper, L., and D. Del Bene. 2016. Transforming Knowledge Creation for Environmental and Epistemic Justice. *Current Opinion in Environmental Sustainability* 20: 41–49.
Turner, D.A. 2006. *This Is Not a Peace Pipe: Towards a Critical Indigenous Philosophy*. University of Toronto Press.
Vila Benites, G., and C. Bonelli, eds. 2017. *A contracorriente: agua y conflicto en América Latina*. Quito: Justicia Hídrica, Abya Yala.
Walia, H. 2012. Moving Beyond a Politics of Solidarity Toward a Practice of Decolonization. In *Organize! Building from the Local for Global Justice*, ed. E. Shragge, J. Hanley, and A.A. Choudry. Oakland: PM Press. epub 345–363.
Walker, G. 2012. *Environmental Justice: Concepts, Evidence and Politics*. London: Routledge.
Walsh, C. 2010. Development as Buen Vivir: Institutional Arrangements and (de) colonial Entanglements. *Development* 53 (1): 15–21.
Whyte, K.P. 2017. The Recognition Paradigm of Environmental Injustice. In *The Routledge Handbook of Environmental Justice*, ed. R. Holifield, J. Chakraborty, and G. Walker, 113–123. London: Routledge.
Winter, C.J. 2021. *Subjects of Intergenerational Justice: Indigenous Philosophy, the Environment and Relationships*. London: Routledge.
Yaka, Ö. 2019. Rethinking Justice: Struggles for Environmental Commons and the Notion of Socio-Ecological Justice. *Antipode* 51 (1): 353–372.
Yazzie, M.K. 2018. Decolonizing Development in Diné Bikeyah: Resource Extraction, Anti-Capitalism, and Relational Futures. *Environment and Society* 9 (1): 25–39.
Young, I.M. 1990. *Justice and the Politics of Difference*. Princeton University Press.
———. 2000. Hybrid Democracy: Iroquois Federalism and the Postcolonial Project. In *Political Theory and the Rights of Indigenous Peoples*, ed. D. Ivison, P. Patton, and W. Sanders, 237–258. London; New York: Cambridge University Press.
Zografos, C., and J. Martínez-Alier. 2009. The Politics of Landscape Value: A Case Study of Wind Farm Conflict in Rural Catalonia. *Environment and Planning A: Economy and Space* 41 (7): 1726–1744.
Zwarteveen, M.Z., and R. Boelens. 2014. Defining, Researching and Struggling for Water Justice: Some Conceptual Building Blocks for Research and Action. *Water International* 39 (2): 143–158.

Appendix: Energy Collectives

This appendix illustrates two different models of collective ownership and participation within localized energy initiatives. One is located in Gujarat state in India, the other on the small island of Eigg, off the west coast of Scotland. These two locations are very geographically distant one from the other—and each project takes shape through distinct social histories, economic realities, cultural identities, and political challenges. Even across these important differences, both cases reflect how community-led processes of organization can confront overlapping vulnerabilities that emerge within current social and economic structures. Experiences in both places also show how collective action and diverse local resources can be reconfigured in order to compensate for (or to remove) those vulnerabilities.

All too often, alternative ways-of-life and modes of social organizing that aim at restoring socioecological relations are suppressed by those in positions of power who deem such alternatives to be running counter to national interests or against established ideas of progress.[1] Despite such

---

[1] Blaser, M., 2004. Life Projects: Indigenous Peoples' Agency and Development. *In*: M. Blaser, H. Feit, and G. McRae, eds. *In the Way of Development: Indigenous Peoples, Life Projects and Globalisation*. London: Zed Books;

Partridge, T., 2017. Unconventional Action and Community Control: Rerouting Dependencies Despite the Hydrocarbon Economy. *In*: K. Jalbert, A. Willow, D. Casagrande, and S. Paladino, eds. *ExtrACTION: Impacts, Engagements and Alternative Futures.* New York: Routledge, 198–210.

pressures, the two projects described here show how energy initiatives can support other notions of progress and visions for more just and sustainable futures. In part, this is done by pushing against the apparent inevitability of fossil fuel dependency. Another major component is that such restorative initiatives both draw on and foster community collaboration and cooperation—collective efforts that inevitably bring with them their own challenges, conflicts, and tensions. Part of the success of these initiatives, it should be noted, is due to participants incorporating these challenges into the work of the energy project itself: novel energy relations are embedded within their specific political and ecological contexts.

These are not blueprints. What works in one context might not be appropriate in another. But what these energy collectives underline is the importance and viability of radical social innovations[2]—not relying on high-cost, resource-intensive, corporate-controlled, techno-fix 'innovations' to save us and instead collaboratively building locally-attuned practices of care, restoration, respect, and accountability. These already-existing initiatives embrace many of the critical concerns raised throughout this book: taking seriously the need for radical change across interconnected justice concerns.

[2] Apostolopoulou, E., Bormpoudakis, D., Chatzipavlidis, A., Cortés Vázquez, J.J., Florea, I., Gearey, M., Levy, J., Loginova, J., Ordner, J., Partridge, T., Pizarro, A., Rhoades, H., Symons, K., Veríssimo, C., and Wahby, N., 2022. Radical Social Innovations and the Spatialities of Grassroots Activism: Navigating Pathways for Tackling Inequality and Reinventing the Commons. *Journal of Political Ecology*, 29 (1), 144–188.

I. Dhundi Saur Urja Utpadak Sahakari Mandali [Dhundi Solar Pump Irrigators' Cooperative Enterprise] (Gujarat, India)

Image 1 Dhundi Saur Urja Utpadak Sahakari Mandali (drying site i). (Photo © Tristan Partridge, 2018)

Image 2   Dhundi Saur Urja Utpadak Sahakari Mandali (drying site ii). (Photo ©
Tristan Partridge, 2018)

Image 3   Dhundi Saur Urja Utpadak Sahakari Mandali (drying field i). (Photo ©
Tristan Partridge, 2018)

Image 4   Dhundi Saur Urja Utpadak Sahakari Mandali (drying field ii). (Photo © Tristan Partridge, 2018)

Image 5   Fudabhai Parmar (Dhundi, Gujarat). (Photo © Tristan Partridge, 2018)

Image 6 Fudabhai Parmar's solar installation (Dhundi, Gujarat). (Photo ©
Tristan Partridge, 2018)

Image 7   Still from film "Dhundi Saur Urja Utpadak Sahakari Mandali" (i) https://www.youtube.com/watch?v=iUZYShkPdQg. (© Tristan Partridge, 2020)

Image 8 Still from film "Dhundi Saur Urja Utpadak Sahakari Mandali" (ii) https://www.youtube.com/watch?v=iUZYShkPdQg. (© Tristan Partridge, 2020)

Dhundi 'Saur Urja Utpadak Sahakari Mandali' (Solar Pump Irrigators' Cooperative Enterprise)

In the state of Gujarat, India, the village of Dhundi was selected as the site for what is claimed to be the world's first 'solar cooperative that produces Solar Power as a Remunerative Crop'—a combined approach to localizing solar energy generation, addressing groundwater management, and supplementing farmer incomes, since dubbed the SPaRC model.[3] The Dhundi "Solar Pump Irrigators' Cooperative Enterprise"—the acronym spells SPICE—began in May 2016 with investment and support from the International Water Management Institute (IWMI), part of a global research network which has offices in the nearby district capital, Anand.

The Dhundi solar cooperative uses solar energy to run irrigation water pumps—replacing diesel-fueled water pumps—and to generate income by selling surplus power to the grid. Across India, the scale and number of solar installations have seen rapid recent growth and continue to grow. Solar-powered water pumps are no exception to this trend. National estimates track the number of solar pumps at less than 7500 in 2010 up to almost 100,000 in 2016–2017.[4] As IWMI project coordinators emphasize, however, often solar-powered water pumps continue to run whether or not farmers need (or are able to use) the power directly for irrigation, since usually there is no opportunity to sell any 'surplus' solar energy. That surplus is thus effectively wasted. The Dhundi SPICE project operates differently: farmers can complete (or selectively limit) their irrigation pumping and then, together, pool 'surplus' solar energy and sell it to the grid under a 25-year power purchase agreement with Madhya Gujarat Vij Company Limited (MGVCL), the local power distribution company (or DISCOM).[5]

The history of farmer initiatives in Gujarat is particularly relevant for understanding the success of the SPICE project. Anand, the district capital, is the same city that is home to the Amul dairy group—one of the world's largest cooperatives with millions of members, frequently held up as a prime example of cooperative success. Often this success is framed in terms of regional economics and growth: a mainstay of India's so-called "white revolution" that has grown since its start in the 1940s and has seen India become the world's largest milk producer, Amul is itself now the

---

[3] Shah, T., Durga, N., Rai, G.P., Verma, S., and Rathod, R., 2017. Promoting Solar Power as a Remunerative Crop. *Economic & Political Weekly*, 52 (45), 14–19.

[4] IANS, 2017. Solar Powered Solution for Groundwater Crisis. *The Hans India* (6 June).

[5] Shah et al. (2017).

largest food brand in India.[6] The focus on growth is unsustainable and the model as a whole is far from perfect, but the importance of Amul can also be assessed in more qualitative terms, particularly with regard to its influence on community organizing and collaboration.

Fudabhai Parmar, pictured above cleaning dust from the solar panels that power his groundwater irrigation pump [Image 5], has been a member of the Amul coop for decades. During my fieldwork in Dhundi in 2018, he described how, with his neighbors—when the solar cooperative project was proposed—they saw an opportunity to both save money on diesel fuel and to make money by selling solar energy to the grid.

Amul's influence in the region is clear: cooperative organizing, joint ownership, collective earning, and pooling resources for mutual economic benefit—these are practices that, rather than being dismissed or viewed with suspicion, are widely seen as positive and productive modes of organizing.

The solar cooperative has been presented by IWMI and its funders as a 'proof of concept' for promoting the SPaRC (Solar Power as a Remunerative Crop) model: the idea being that farmers can 'grow' and sell solar energy as a cash crop—a crop, they also point out, that "needs no seeds, fertiliser, pesticides, irrigation or backbreaking labour; and which has a ready buyer at their door-step at an assured price: income from the solar 'crop' is free of risk from droughts, floods, pests and diseases".[7]

The farmers I spoke with pointed out that most of their time is still dedicated to tending crops (particularly maize, sugarcane, mangos, and pearl millet) and also to managing irrigation (such as maintaining canals and bunds). Not only have solar pumps replaced diesel pumps in Dhundi; the solar system also offers a more stable energy supply than the state subsidized grid (which runs only for 7 or 8 hours per day, frequently cuts out, and is supplied during the night for half the days in a month). Irrigating during the day with solar is much less hazardous and wasteful.

Solar power is also cheaper, once the panels have been installed. While the first six members of the Dhundi solar cooperative contributed less than 10% of set-up costs, 4 farming households have joined since then, and were willing to pay close to 40%. The cooperative has also been designed to reduce the amount of groundwater being pumped and exploited: the

[6] UN-INDIA, n.d. *UNICEF Signs Agreement with Government of India to Fund the Aarey and Anand led White Revolution.*

[7] Shah et al. (2017, p. 15).

idea here is that selling surplus power back to the grid incentivizes farmers to minimize the amount of energy they use for pumping water (and therefore to maximize the amount of solar energy they can sell).

The goals, then, of the International Water Management Institute are broad in scope: "to reconfigure our power economy, our groundwater economy, and our agrarian livelihoods".[8] However, beyond any financial incentives that might make the solar project attractive to farming households, its ongoing success is dependent upon community labor, collaboration, and specific social resources—perhaps most notably a particular regional history of previous community engagements with agricultural cooperatives and their benefits.

A core goal of India's National Solar Mission is to create 100GW of solar capacity by the end of 2022; as of February 2022, total installed solar capacity was 50GW nationwide.[9] Solar energy generation has also been associated with the government pledge to double all farmers' incomes, also by the end of 2022. Even though that claim (and any progress toward achieving it) has dropped out of government rhetoric—the February 2022 Union Budget address made no mention of it[10]—the apparent promise of decentralized energy initiatives (and of solar power in particular) to help achieve these targets increasingly informs policy and decision-making. Government agencies across India are turning to the technical deployment of solar power as a means to address a range of social issues.[11] Every aspect of these plans presents challenges and questions, not least concerning how equitable, redistributive, and environmentally sustainable they will be.

At the local scale, however, experiences with the Dhundi SPICE underline something critical to the construction of alternative energy futures: global growth in decentralized solar projects is a phenomenon that still largely depends upon community work, relationships, and cooperation in order to succeed—and such projects benefit enormously when rooted in communities who already have histories of collective action and prior experiences of working collaboratively.

[8] Ibid.

[9] OIES, 2022. *India's Progress on its Climate Action Plan—An Update in Early 2022.* Oxford Institute for Energy Studies.

[10] Vissa, K., 2022. A Grim Future: What Happened to the Promise of Doubling Farmers' Income by 2022? *The Wire* (5 February).

[11] Partridge, T., 2020. "Power Farmers" in North India and New Energy Producers around the World: Three Critical Fields for Multiscalar Research. *Energy Research & Social Science*, 69, 101575.

II. Eigg Electric Ltd. (Scotland, U.K.)

Image 9   Eigg solar array (i). (Photo © Greg Carr, 2022)

Image 10   Eigg solar array (ii). (Photo © Greg Carr, 2022)

Image 11 Still from film "Eigg: Reclaimed and Renewable". (© Tristan Partridge, 2014)

Image 12 Still from film "Eigg: Reclaimed and Renewable". (© Tristan Partridge, 2014)

Image 13   Eigg solar array (and An Sgùrr). (Photo © Greg Carr, 2022)

Image 14   Eigg solar array (and view beyond). (Photo © Greg Carr, 2022)

Eigg, Land Reform, and Decarbonization[12]

The Isle of Eigg is a small island located 10 miles off the west coast of Scotland (population c.100). There has never been any connection between the island and the mainland electricity grid. In 2008, an island-wide electricity micro-grid based on renewable sources was completed, altering how residents use and depend upon fossil fuels. Reclaiming community ownership of the land has been a vital component of making this energy initiative an ongoing success.

In June 1997 the residents of Eigg purchased the island they called home, bringing an end to years of insecure relations between the 60–70 residents and the individual who had been the sole owner and landlord. In collaboration with supporters and activists from other highland and island communities in Scotland, as well as with financial support from thousands of donations, a community Trust was formed to buy out the previous owner. Following the buy-out, a residents' association and the Trust board members were to have full decision-making powers: their vision was to establish community control over island planning, infrastructure, and development—effectively to guide the future of Eigg.[13]

Eigg was not the first example in the region of collective organizing to localize governance. In Assynt, local crofters who were farming the land set an important precedent by forming a Trust in 1992. The following year, they took a privately-owned estate into community ownership by mounting an 'open market bid' with outside financial support—the same model that was adapted by Eigg in 1997, and subsequently Knoydart in 1999 and Gigha in 2002.[14] Community buy-outs emerged as a Scottish land reform movement, but one that preceded institutional

---

[12] This text is a modified version of a previous publication, used with permission. Source: Partridge, T., 2017. Unconventional Action and Community Control: Rerouting Dependencies Despite the Hydrocarbon Economy. *In*: K. Jalbert, A. Willow, D. Casagrande, and S. Paladino, eds. *ExtrACTION: Impacts, Engagements and Alternative Futures.* New York: Routledge, pp. 198–210.

[13] Braid, M., 1996. Between a Rock and a Hard Place. *The Independent* (26 February);
McIntosh, A., 2003. The Precious Burden: 4 Stages of Land Reform—Scotland. *The Hebridean* (2 October).

[14] Hunter, J., Peacock, P., Wightman, A., and Foxley, M., 2013. *432:50—Towards a Comprehensive Land Reform Agenda for Scotland.* Scottish Affairs Committee: Parliament of the United Kingdom.

change—Scotland's official Land Reform Act wasn't ratified until 2003.[15] It is a movement that has been pushing back against drastic inequalities in land ownership. For instance, in 2009, just 0.025% of Scotland's population owned 67% of privately-held rural land.[16] Daniel Rhys Morgan's ethnographic analysis of the Eigg buy-out provides a range of insights into the process. Two aspects are underscored: the influence of the Assynt success; and Eigg's innovative strategy of building a partnership to purchase and manage the island, whereby islanders joined with the Scottish Wildlife Trust and the regional Highland Council.[17] Cases like Eigg illustrate how localized action has initiated political changes both within and beyond local legal structures.

Today, islanders participate in the partnership via the Eigg Residents' Association (ERA). Four ERA directors, elected by the community, meet every three weeks with responsibility for day-to-day management. In addition, there are three subsidiary companies of the Isle of Eigg Heritage Trust (Eigg Trading Limited; Eigg Construction Limited; Eigg Electric Ltd.) each with a board of directors appointed by community-members.[18] The emphasis on management here is critical: collectively designing and governing island infrastructure has been key to ensuring that the Eigg micro-grid electrification project was not only completed but completed in such a way that served future stability and sustainability goals.

As a small island (five miles by three) with a population of about 100 (and growing since the buy-out), planners found that the costs involved in potentially linking Eigg to the mainland electric grid were prohibitive. For many years, electricity on Eigg came from individually owned diesel-fuel generators dotted across the landscape. Following the buy-out, plans for an alternative system were developed between islanders, renewable energy specialists, and outside technicians with support from public agencies and charitable organizations. In 2008, following extensive community campaigning and fundraising, construction was completed on an island-wide electricity micro-grid based on renewable sources. The old household

---

[15] p. 195: Kenrick, J., 2011. Scottish Land Reform and Indigenous Peoples' Rights: Self-determination and Historical Reversibility. *Social Anthropology/Anthropologie Sociale*, 19 (2), 189–203.

[16] Warren, C., 2009. *Managing Scotland's Environment*. Edinburgh University Press.

[17] pp. 13–14: Morgan, D., 1999. *The Isle of Eigg: Land Reform, People, and Power*. Ph.D. Thesis. University of Edinburgh.

[18] IEHT, n.d. *The Structure of the Isle of Eigg Heritage Trust: Subsidiary Companies*. Isle of Eigg Heritage Trust.

generators were replaced as the primary source of electricity and, as one resident put it (during my fieldwork on Eigg in 2014), "we heard silence again."

Electricity now comes from three hydroelectric generators (totaling up to 119kW), four 6kW wind turbines, and a 30kW photovoltaic solar array that has since been expanded to 50kW, angled at 20 degrees to the horizontal—with two 64kW diesel generators for use in emergencies, during maintenance, or to meet demand peaks. Storage and distribution are controlled via an array of ninety-six 4volt batteries. Island estimates put the supply from renewables at about 95%.[19] Residents have also found this new energy supply more reliable and sustained, and less dependent on the delivery and fluctuating prices of barrels of diesel sourced from the mainland. This transition not only represents a vital shift in how energy needs are met on Eigg, it has also strengthened the island's reputation as a leader in local and national environmental action.

Transitioning away from diesel generators to a localized micro-grid based on renewable energy has not created isolation, however. Electricity may now be island-based, and more appropriate to the needs of residents, but its generation still requires maintenance components and expertise sourced from different parts of the UK and the world, together with small amounts of backup diesel. Daily life is also largely dependent on medical, food, and other provisions that are ferried in from off-island locations. These are some of many examples where communities—including relatively remote or island community such as Eigg—remain deeply linked with distant places and projects. Entanglements with the dependencies of fossil fuel extraction—and on products derived from those fossil fuels—are still present. By creating the micro-grid, relations of dependence on fossil fuels were lessened and redefined but not entirely severed.

Rearranged dependencies can also come as a result of collective reassessments of use and need. For example, at the design stages, the electrification project also specifically addressed patterns of consumption. Generator maintenance on older properties had revealed disproportionate levels of electricity use across households of a similar size, due mainly to aging and inefficient appliances—a problem planners felt the new grid could solve by regulating or 'sharing' available supply, effectively limiting the number of appliances that can be used at any one time. Residents and project designers calculated a 5kW-per-household limit (a higher limit of

---

[19] Source: http://isleofeigg.org/eigg-electric/ Accessed: 6 April 2022.

10kW is set for commercial properties). Building the micro-grid from scratch made it possible to install "OWL meters," electricity monitors that track usage and govern the power limit at each property. The 5kW limit also made it cheaper to lay cabling for the micro-grid itself; a total of 11km of underground cable constitutes the distribution network of the micro-grid. Aluminium wire could be used in place of more expensive copper. This measure saved on construction costs, while the OWL meters provide users with live tracking of energy usage: residents can monitor, manage, and reflect upon their personal dependencies on electricity consumption. Together, these specific modes of generating and delivering electricity to island residents invite critical reassessment of the island's energy use at both individual and community-wide scales.

Eigg's story of re-assigning property ownership through land reform has had consequences beyond those quantified in economic terms and these changes continue to enable the pursuit of an àlternative energy future. Deeper personal investment of time, money, and effort in these projects has increased motivation and support for future actions—since the people involved have been exposed in new ways to the relationships (and interdependencies) that connect them with others and with the island, its needs, and its operations as a whole. These dynamics are witnessed elsewhere in the world. If, as McIntosh argues, land reform "empowers people to settle more deeply into their geographical place" and enables people "to make decisions about the habitation, resource use and amenity enrichment of their own place,[20]" it does so by creating structures of leadership and community governance that ensure democratic accountability. Changes to how people on Eigg now access electricity are intricately bound to projects and possibilities that were enabled by the community buy-out—by meaningful land reform.

Rather than offering a replicable model for change, the recent history of Eigg reflects an understanding of resilience and pathways to change that is geographically and socioculturally specific. For many communities, organizing to create alternative futures requires developing localized capacities to withstand and respond to unwanted change often driven by external influences and events. Rennie and Billing note that this resilience is found by "utilizing the human and natural resources of [an] area in a manner that sensitively exploits the ability of these resources to adapt to

---

[20]p. 3: McIntosh, A., 2015. *Consultation Response to the Future of Land Reform in Scotland.* Govan: Centre for Human Ecology.

and benefit from change".[21] This understanding of resilience creates a basis for alterative futures that are locally relevant, appropriate, and responsive.

The decarbonization of electricity supplies on Eigg involved actions embedded within other collective projects. These actions were enabled by three key factors, each reinforcing one another: (i) collective decision-making, (ii) direct control of land and other resources, and (iii) forms of social organizing developed in dialogue with similar communities elsewhere. These processes illustrate how inseparable energy justice concerns are from related environmental justice needs and broader social struggles.

[21] p. 35: Rennie, F. and Billing, S.-L., 2015. Changing Community Perceptions of Sustainable Rural Development in Scotland. *Journal of Rural and Community Development*, 10 (2), 35–46.

Index[1]

---

[1] Note: Page numbers followed by 'n' refer to notes.